04-WEM-4

Evaluating Waterbody Assessment and Listing Processes: Integration of Monitoring and Evaluative Techniques

by:

Bret Linenfelser
City of Boulder

Lindsay Griffith
Brown and Caldwell

2007

WERF

The Water Environment Research Foundation, a not-for-profit organization, funds and manages water quality research for its subscribers through a diverse public-private partnership between municipal utilities, corporations, academia, industry, and the federal government. WERF subscribers include municipal and regional water and wastewater utilities, industrial corporations, environmental engineering firms, and others that share a commitment to cost-effective water quality solutions. WERF is dedicated to advancing science and technology addressing water quality issues as they impact water resources, the atmosphere, the lands, and quality of life.

For more information, contact:
Water Environment Research Foundation
635 Slaters Lane, Suite 300
Alexandria, VA 22314-1177
Tel: (703) 684-2470
Fax: (703) 299-0742
www.werf.org
werf@werf.org

This report was co-published by the following organization. For non-subscriber sales information, contact:

IWA Publishing
Alliance House, 12 Caxton Street
London SW1H 0QS, United Kingdom
Tel: +44 (0) 20 7654 5500
Fax: +44 (0) 20 7654 5555
www.iwapublishing.com
publications@iwap.co.uk

Library of Congress Catalog Card Number: 2006931055
IWAP ISBN: 978-1-84339-774-8

Brown and Caldwell

The research on which this report is based was developed, in part, by the United States Environmental Protection Agency (EPA) through Cooperative Agreement No. CR831559-01 with the Water Environment Research Foundation (WERF). However, the views expressed in this document are solely those of Brown and Caldwell and neither EPA nor WERF endorses any products or commercial services mentioned in this publication. This report is a publication of WERF, not EPA. Funds awarded under the Cooperative Agreement cited above were not used for editorial services, reproduction, printing, or distribution.

ACKNOWLEDGMENTS

The Research Team would like to thank those organizations and individuals who provided their time and effort to review this project.

Report Preparation

Principal Investigator:

Bret Linenfelser
City of Boulder, CO (formerly with Brown and Caldwell)

Lindsay Griffith
Brown and Caldwell

Project Team:

Nadine Adkins
Lavanya Gullapalli
Brown and Caldwell

Expert Panel:

Fredric Andes
Barnes & Thornburg LLP

Dennis Helsel
U.S. Geological Survey

Susan Holdsworth
U.S. Environmental Protection Agency

Tim Moore
Risk Sciences

Project Subcommittee

Pixie A. Hamilton
U.S. Geological Survey

Heather Lamberson
Los Angeles County Sanitary Districts (LACSD)

Jim Pendergast
Research Council Liaison – U.S. Environmental Protection Agency

Irwin Polls
Metro Water Reclamation District of Greater Chicago (Retired)

Tom Stiles
Kansas Department of Health and Environment

Robert Ward
Colorado State University (Retired)

Water Environment Research Foundation Staff

Director of Research: Daniel M. Woltering, Ph.D.
Program Manager: Jane M. Casteline

ABSTRACT AND BENEFITS

Abstract:

Under the Clean Water Act (CWA), state governments are required to prepare comprehensive biennial water quality assessment reports (305(b) reports) and to submit biennial listings of impaired waters (303(d) lists) requiring a Total Maximum Daily Load (TMDL). Each state has developed its own waterbody assessment and listing methodology, resulting in a variety of approaches that are not uniform or standardized. In order to facilitate more effective integration of monitoring data and technically sound assessment methodologies into the waterbody assessment and listing process, WERF provided funding to document and evaluate current assessment and listing methodologies employed by the states during the development of their Integrated Reports.

Research tasks included gathering information on current assessment methodologies, and identifying those methodologies that optimize available data and characterize waterbody conditions with the greatest confidence. Despite improvements over the last decade in both monitoring and assessment, the states continue to struggle with the uncertainty involved with determining attainment of designated uses. In order to improve the assessment and listing process, the research team recommends that states: 1) develop and publish minimum data requirements, 2) better integrate monitoring programs with waterbody assessment needs, 3) develop standardized assessment units that allow for water quality extrapolation, 4) develop numeric water quality criteria by which to more reliably determine support or impairment of designated uses, 5) use statistically based data evaluation techniques to more confidently determine attainment of water quality standards, and 6) include the public in the methodology development process.

Benefits:

- Documents the current "state-of-the-science" regarding the waterbody assessment process employed by the states.
- Provides guidance on integration of monitoring data with analysis methods so that states can better understand waterbody ecological condition.
- Provides guidance on how to evaluate data using scientifically defensible methodologies that can characterize water quality with greater confidence.
- Provides guidance on how to more reliably determine waterbodies which should be designated as impaired.

Keywords: Waterbody assessment, integrated report, water quality monitoring, water quality assessment, assessment methodology.

TABLE OF CONTENTS

Acknowledgments iii
Abstract and Benefits v
List of Tables viii
List of Figures ix
List of Acronyms x
Executive Summary ES-1

1.0 Introduction 1-1
1.1 Project Rationale and Background 1-1
1.2 Purpose of Research 1-3
1.3 Project Objectives 1-3
1.4 Project Timing 1-4

2.0 Overview of Research Tasks 2-1
2.1 Research Team and Expert Panel 2-1
2.2 Literature Review of States' Integrated Reports 2-3
2.3 Interviews with State Personnel 2-3
2.4 Development of Critical Evaluation Matrix 2-4
2.5 Review of Assessment Guidance Documents 2-4
2.6 Development of Recommendations 2-5

3.0 Research Findings: What Do Current Waterbody Assessments Look Like? 3-1
3.1 U.S. EPA's Integrated Report Guidance 3-1
3.2 What Specific Types of Data Are Required to Conduct Waterbody Assessment Methodology for Listing Purposes? 3-2
3.2.1 Data Sufficiency 3-2
3.2.2 Data Quality Requirements 3-4
3.3 How Is the Assessment Methodology Tied to Monitoring Network Design? 3-5
3.4 How Does Assessment Methodology Extrapolate Water Quality over Space/Time Where Data Are Not Available? 3-6
3.5 How Does Assessment Methodology Specifically Determine Water Quality Standard Attainment or Non-Attainment? 3-6
3.6 Does Assessment Methodology Quantify Uncertainty? 3-11
3.7 Was Public Participation/Peer Review Sought When Developing Assessment Methodology? 3-12

4.0 Recommendations 4-1
4.1 Data Requirements to Conduct Waterbody Assessment Methods for Listing Purposes 4-2
4.1.1 Develop and Publish Minimum Data Quality Requirements 4-2
4.1.2 Describe How Waterbodies Will Be Assessed if Data Quality Requirements Are Not Met 4-3
4.1.3 Address Data Censorship Issues in Assessment Methodology 4-3
4.2 How the Assessment Method Is Tied To Monitoring Network Design 4-4

4.2.1 Integrate Statewide Assessment Monitoring with Waterbody-Specific Assessment Efforts 4-4
4.2.2 Allocate State Resources to Long-Term Monitoring on Watershed and Site-Specific Scales 4-5
4.2.3 Review Monitoring Objectives from Outside Data Sources 4-6
4.2.4 Develop Biological Monitoring and Assessment Program to Support Biocriteria Development 4-7
4.2.5 Consider Emerging Evaluative Techniques to Analyze Data from Multiple Sources 4-7
4.3 Water Quality Extrapolation over Space/Time Where Data Are Not Available 4-9
4.3.1 Geo-Reference Water Quality Monitoring Data 4-9
4.3.2 Flow-Reference Water Quality Data 4-9
4.3.3 Standardize the Development of Assessment Units 4-9
4.4 How States Determine Water Quality Standard Attainment 4-10
4.4.1 Develop Numeric Criteria to Translate Narrative Water Quality Standards 4-10
4.4.2 Allow for Weight-of-Evidence Approach to Determine Both Attainment and Non-Attainment of Beneficial Uses 4-12
4.4.3 Use Statistically Based Evaluative Techniques for Comparing Chemical Data to Numeric Water Quality Criteria 4-14
4.4.4 Base Determination of Fully Supporting On Attainment of Minimum Water Quality 4-15
4.4.5 Develop Better Descriptions of the Linkage Between 305(b) and 303(d) Waterbody Assessments 4-16
4.4.6 Provide More Detailed Descriptions of Assessment Methodologies to the Public 4-16
4.4.7 Provide Descriptions of Specific De-Listing Assessment Methodologies 4-17
4.5 Public Participation/Peer Review During Methodology Development 4-17
4.5.1 Provide a Forum for Public Input During Development of Waterbody Assessment Methodology 4-17
4.5.2 Submit Draft Assessment and Listing Methodology Document to U.S. EPA for Review and Comment 4-17

5.0 Summary and Conclusions 5-1
5.1 What Do Current Waterbody Assessment Methods Look Like? 5-1
5.2 What Does a Model Waterbody Assessment Methodology Look Like? 5-2
5.2.1 Data Used in Waterbody Assessments 5-3
5.2.2 Integrating Monitoring with Assessments 5-3
5.2.3 Water Quality Extrapolation Techniques 5-3
5.2.4 Determining Water Quality Standards Attainment 5-4
5.2.5 Public Involvement 5-4

Appendix A: Report Search Inventory A-1
Appendix B: Template Evaluation Survey B-1
Appendix C: State Evaluation Matrix C-1
Glossary of Terms G-1
References R-1

LIST OF TABLES

3-1 U.S. EPA Integrated Report Five Reporting Categories 3-1
3-2 Core Indicators Used by the State of Arizona to Determine Attainment Status of Waters 3-8
4-1 Independent Applicability For Purposes of Water Quality Standard Attainment/Nonattainment Determinations 4-13

LIST OF FIGURES

2-1 Summary of Study Approach Taken by Research Team ... 2-2
3-1 States With Minimum Sample Size Requirements ... 3-3
3-2 States Using the Ten-Percent Rule to Determine Compliance with Specific Water Quality Standards ... 3-9
3-3 States Which Use Weight-of-Evidence Approach to Determine Use Support ... 3-11
3-4 States Which Quantify Uncertainty in Their Assessments ... 3-12
3-5 States Which Solicit Public Comment During Assessment Methodology Development ... 3-13
4-1 Benefits of Developing and Publishing Minimum Data Requirements ... 4-2
4-2 Benefits of Allocating State Resources to Long-Term Monitoring or Watershed and Site-Specific Scales ... 4-6
4-3 Benefits of Developing Biological Monitoring and Assessment Program to Support Biocriteria Development ... 4-7
4-4 Benefits of Using Flow-Reference Water Quality Data ... 4-9

LIST OF ACRONYMS

ADB	Assessment Database
ADEQ	Arizona Department of Environmental Quality
AU	Assessment Unit
CWA	Clean Water Act
EMAP	Environmental Monitoring and Assessment Program
GIS	Geographic Information System
HUC	Hydrologic Unit Code
IBI	Index of Biotic Integrity
MLE	Maximum Likelihood Estimation
NHD	National Hydrography Dataset
NPDES	National Pollutant Discharge Elimination System
NRC	National Research Council
PSC	Project Subcommittee
QA/QC	Quality Assurance and Quality Control
QAPP	Quality Assurance Project Plan
SSC	Suspended Sediment Concentration
TMDL	Total Maximum Daily Load
U.S. EPA	United States Environmental Protection Agency
USGS	United States Geological Survey
WERF	Water Environment Research Foundation

EXECUTIVE SUMMARY

To characterize water quality conditions for waterbodies within the state, as well as to fulfill Clean Water Act reporting requirements under sections 305(b) and 303(d), the state governments (states) have developed waterbody assessment and listing methodologies which aim to evaluate current water quality conditions and determine if these conditions meet water quality standards. Waterbodies which do not meet water quality standards are listed on the state's 303(d) list of impaired waters. Limitations in monitoring and assessment resources available to the states, coupled with the uncertainty inherent in the waterbody assessment and listing process, serve to produce information which may not adequately characterize water quality conditions nor allow states to confidently determine support or non-support of designated uses. In order to facilitate more effective integration of monitoring data and technically sound assessment methodologies into the state waterbody assessment and listing process, a critical evaluation of current methodologies employed by the states is warranted.

In August 2005, the research team was selected by the Water Environment Research Foundation (WERF) to evaluate current assessment and listing procedures used by states and to develop peer-reviewed guidance on evaluative methods that will reduce uncertainty in the waterbody assessment and listing process.

Specific research tasks included the following:

1. Gather information on current methodologies that the states and others use to assess waterbody conditions for listing purposes.
2. Identify current assessment approaches which optimize available monitoring data and best characterize waterbody conditions and associated uncertainty.
3. Recommend assessment methodologies to reliably determine attainment and non-attainment of designated uses.

To ensure a basic understanding of each state's assessment and listing methodology, and to facilitate comparison of the strengths and weaknesses of each methodology, the following questions were considered:

1. What types of data are required to conduct waterbody assessment methodology for listing purposes?
2. Are there quality assurance/quality control (QA/QC) requirements for data used in the assessment?
3. If data do not meet QA/QC requirements, how are the data used in the assessment?
4. How is the assessment methodology tied to monitoring network design?
5. How does the assessment methodology extrapolate water quality over space and time where data are not available?
6. How does the assessment methodology determine water quality standard attainment or non-attainment?

7. How does the assessment methodology quantify uncertainty?
8. Was public participation and/or peer review sought when developing the assessment methodology?

The research team compiled and summarized information regarding the states' assessment and listing methodologies, focusing on the most common methodologies and unique methodologies with particular strengths for determining support of designated uses with consistency and confidence. Despite vast improvements by the states over the last decade in both monitoring and assessment, the states continue to struggle with the uncertainty involve with determining attainment of designated uses. For many states, the biannual assessment and listing cycle continues to evolve as they try to determine which monitoring and assessment methodologies, utilizing chemical or biological water quality indicators, or both, should be the focus of their limited funding and resources.

Using information from guidance documents, expertise of team members, and current state assessment methodologies, the research team developed recommendations to assist the states to develop and/or refine their waterbody assessment and listing methodologies. The recommendations are organized under the key methodology components and are summarized below.

1. Data Used in Waterbody Assessments

- Develop and publish minimum data quality requirements
- Specifically describe how waterbodies will be assessed if data quality requirements are not met
- Address data censorship issues in the assessment methodology

2. Integrating Monitoring with Assessments

- Integrate statewide assessment monitoring with waterbody-specific assessment efforts
- Allocate state resources to long-term monitoring on watershed and site-specific scales
- Review monitoring objectives for outside data sources
- Design a biological monitoring and assessment program to support biocriteria development
- Consider emerging evaluative techniques for analyzing data from multiple sources

3. Water Quality Extrapolation Techniques

- Geo-reference water quality monitoring data
- Flow-reference water quality monitoring data
- Standardize the development of assessment units

4. Determining Water Quality Standards Attainment

- Develop numeric criteria to translate narrative water quality standards
- Allow for a weight-of-evidence approach to determine both attainment and non-attainment of beneficial uses
- Use statistically based evaluative techniques to compare chemical data to numeric water quality criteria

- Base determination of "fully supporting" on attainment of minimum water quality
- Develop better descriptions of the linkage between 305(b) and 303(d) waterbody assessments
- Provide more detailed descriptions of assessment methodologies to the public
- Provide descriptions of specific de-listing assessment methodologies

5. Public Involvement

- Provide a forum for public input during development of the waterbody assessment methodology
- Submit a draft assessment and listing methodology document to the U.S. Environmental Protection Agency (U.S. EPA) for review and comment

The recommendations listed above will provide guidance to the states on how to more effectively integrate monitoring program design with better data evaluation techniques and data quality management programs. Several recommendations also address development of assessment methodologies that are more scientifically defensible and that can characterize water quality with greater confidence. Further research on the adequacy of specific data evaluative techniques is needed to provide guidance to the states on the most appropriate methods given the characteristics of their monitoring data. These recommendations are intended to provide a technical resource to the states so that they can determine, with greater reliability, waterbodies that should be designated as impaired and included on their 303(d) lists.

CHAPTER 1.0

INTRODUCTION

1.1 Project Rationale and Background

Under the Clean Water Act (CWA), state governments (states) are required to prepare comprehensive biennial water quality assessment reports (305(b) reports). Additionally, states are required to submit biennial listings of impaired waters (303(d) lists) requiring a Total Maximum Daily Load (TMDL). In recent years, the U.S. EPA has encouraged the states to submit a single waterbody monitoring and assessment report to satisfy both of these requirements (Integrated Report). In order to characterize water quality conditions for waterbodies within the state, as well as to fulfill CWA reporting requirements, the states have developed waterbody assessment and listing methodologies which aim to evaluate current water quality conditions and determine if these conditions meet water quality standards. U.S. EPA describes the assessment methodology as "the decision process (including principles of science, statistics and logic used in interpreting data and information relevant to segment conditions) that a state employs to determine waterbody attainment of water quality standards" (U.S. EPA, 2005a).

In recent years, some members of the water quality and regulatory community have identified a need for more uniform and standardized listing methodologies and criteria. In 2000, a review of state assessment and listing methodologies concluded the following:

> When reviewing state assessment methodologies and other water quality studies, it is evident that the analysis procedures fall short of providing indisputable information due to the fact that the assessments are often based on subjective narrative criteria or relatively small monitoring data sets, and lack broadly peer-reviewed and agreed upon data analysis methods. (Griffith et al., 2001)

A 1999 report by an anonymous group of U.S. EPA and other agency employees was critical of the consistency and comparability of waterbody assessments conducted by states. The report states that "inconsistencies in the amounts of water monitored or evaluated as well as variations in how impairment and designated use attainment are measured, produce a hodgepodge of information that is of little value in determining national water quality trends or comparing water quality among individual states" (PEER, 1999). A 2001 report by the National Research Council (NRC Report) states that:

> Currently, the states use vastly different frameworks for monitoring and assessment, the net result of which is widely divergent estimates of the extent of impaired waters and of the proportion of waters that are fully assessed. This casts a great deal

> of uncertainty not only about what water quality problems are the most important, but also about the accuracy and completeness of their delineation. (NRC, 2001)

The NRC report goes on to recommend that the "EPA should endorse statistical approaches to proper monitoring design, data analysis, and impairment assessment" and that "states should be required to report the statistical properties of the sample data analyses used to make listing determinations" (NRC, 2001).

Since the publication of the above referenced critiques, U.S. EPA has published guidance for the states' waterbody assessment process which encourages the development and documentation of technically sound assessment and listing methodologies, including its *Consolidated Assessment and Listing Methodology* guidance (U.S. EPA, 2002). In response to the Federal Data Quality Act (P.L. 106-544, H.R. 5658), U.S. EPA also has published a series of guidance documents to help the U.S. EPA and the states implement data and information quality management programs (U.S. EPA, 2001, 2006a, 2006b, 2006c). Finally, U.S. EPA has provided guidance with every listing cycle that is specific to the development of the states' Integrated Reports (U.S. EPA 2003a, 2005a). However, this guidance leaves the development of the waterbody assessment and listing methodology to the states and does not outline specific methodologies for the various data types and monitoring objectives utilized by the states in their assessments.

Previous guidance (U.S. EPA, 2003a) stated that U.S. EPA would not "approve nor disapprove the methodology," but only consider the methodology in its review of the states' 303(d) lists. However, the most recent guidance from U.S. EPA, published for the 2006 assessment and listing cycle, asserts that U.S. EPA will consider the state's methodology "to the extent that it reflects a reasonable interpretation of the state's water quality standards and sound science, in determining whether to approve or disapprove the section 303(d) list" (U.S. EPA, 2005a). The 2006 assessment and listing guidance also indicates that the methodology should

> 1) explain how the state identifies, considers (evaluates) all existing and readily available data and information;
>
> 2) articulate the basics of the quality assurance and quality control (QA/QC) criteria used to evaluate data submitted by outside entities to determine what weight, if any, should be assigned to said data and information; and
>
> 3) explain the analytical approaches, including statistical analyses, used to infer true segment conditions from all valid existing and readily available information.
>
> The decision processes the states describe in the methodology should provide all stakeholders with the opportunity to understand exactly how assessment decisions are made. (U.S. EPA, 2005a)

Despite the technical guidance and information produced over the past five years regarding waterbody assessments and data quality management, the states continue to struggle with the uncertainty involved with determining compliance with water quality standards. This uncertainty stems from several factors, including the following:

- States have limited monitoring resources and therefore must rely on data from outside sources. These data are often of unknown quality.
- There are many waterbodies in each state for which no or very limited current monitoring data exist and so assessment must be extrapolated from the monitoring location to nearby waters.
- Outcomes of the assessment methodologies used by the states may be limited by the data used in the analysis, or by the methodology itself. This can make it difficult to reliably determine impairment/non-impairment with a known degree of confidence.

These sources of uncertainty complicate the selection of evaluative techniques appropriate for analyzing water quality data. The limitation in monitoring and assessment resources, coupled with the uncertainty inherent in the waterbody assessment and listing process, serve to produce information which may not adequately characterize water quality conditions or allow states to confidently determine support or non-support of designated uses.

1.2 Purpose of Research

In order to facilitate more effective integration of monitoring data and technically sound assessment methodologies by the states into the waterbody assessment and listing process, a critical evaluation of the current methodologies employed by the states is warranted. U.S. EPA review of the states' Integrated Reports or 303(d) lists is facilitated by submission of the corresponding assessment methodology documentation.

The research team described herein was selected by the Water Environment Research Foundation (WERF) to perform this critical evaluation, and develop peer-reviewed guidance of appropriate evaluative methods for various data sources to improve confidence in the waterbody assessment process. The results of this research should allow refinement of the Integrated Reporting process to produce more reliable and consistent determinations of attainment/non-attainment of water quality standards. More accurate assessment of waterbody health will allow states to identify and prioritize areas of impairment and promote more efficient use of limited data and limited state funds.

1.3 Project Objectives

Specific objectives for the research included:

- Gather information on the current methodologies that regulatory agencies and others use to assess waterbody conditions.
- Identify current assessment approaches which optimize available monitoring data and best characterize waterbody conditions and associated uncertainty.
- Provide recommendations regarding the most appropriate assessment methodologies to determine attainment or non-attainment of designated uses. These recommendations will serve as guidance to the states on how to:
    - Integrate monitoring program design with evaluative techniques so that better information can be produced from monitoring efforts.
    - Evaluate data using scientifically defensible methods that will characterize water quality with greater confidence.

- Determine with greater reliability waterbodies which should be designated as impaired and included on the states' 303(d) list.

1.4 Project Timing

This research project, which began in August 2005, was conducted during the interim between the states' 2004 and 2006 Integrated Reports. Therefore, the review of the most recent documentation of the states' assessment methodologies is based primarily on the states' 2004 Integrated Report and 303(d) assessment and listing methodology. During the project period, U.S. EPA published its guidance for the 2006 assessment and listing cycle (U.S. EPA 2005a), and modifications to states' assessment methodologies are likely due to changes in recommendations included in U.S. EPA's guidance. The research team has integrated information from U.S. EPA's 2006 assessment and listing guidance into this review, and also attempted to compile information on the most recent methodologies employed by the states, where possible within the time constraints of the research schedule. Although the states' assessment methodologies continue to evolve in response to new guidance and emerging research and technologies, the research team feels that the recommendations identified in this report are more general and programmatic in scope and, based on issues identified in U.S. EPA's 2005 guidance, will apply to the states' 2006 Integrated Reports as well.

CHAPTER 2.0

OVERVIEW OF RESEARCH TASKS

2.1 Research Team and Expert Panel

The research team was headed by *Brown and Caldwell*, an environmental engineering and consulting firm located in Golden, Colorado. *Brown and Caldwell* performed the literature review and other information gathering tasks, as well as the critical evaluation of current waterbody assessment processes. To help in the development of recommendations, the research team also assembled a group of experts in the fields of water quality management and water quality data analysis to serve on an Expert Review Panel for the research project. This Expert Review Panel reviewed the draft report along with the WERF Project Subcommittee (PSC) and provided feedback to ensure that the recommendations from the project can be practically implemented by the states. The Expert Review Panel members represent a broad range of interests and stakeholder groups (i.e., consulting, federal agencies, regulated community), and they have served as a sounding board regarding the scientific validity, practicality and usefulness of the recommendations developed as part of this research effort.

The study approach taken by the research team, summarized in Figure 2-1, is described in the sections that follow.

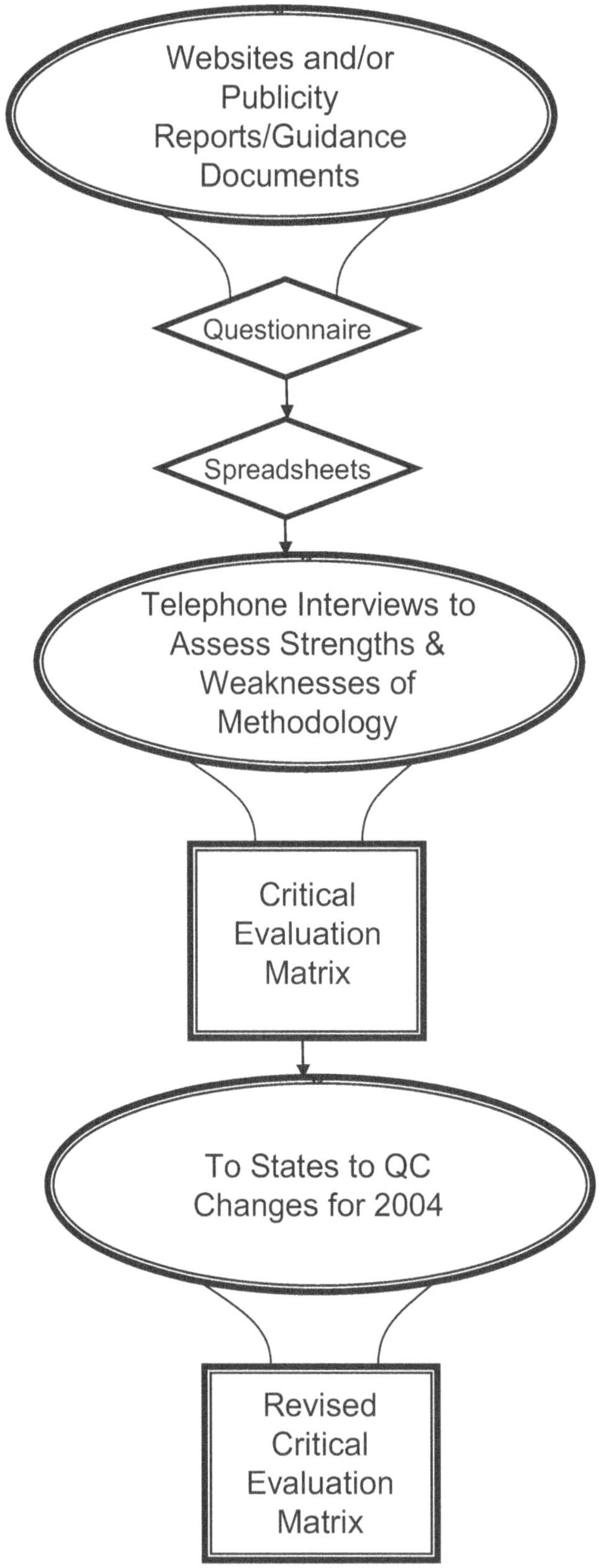

Figure 2 1. Summary of Study Approach Taken by Research Team.

2.2 Literature Review of State Integrated Reports

The first step in the evaluation of waterbody assessment and listing processes was to document the current waterbody assessment methodologies employed by the states during development of their Integrated Reports. To efficiently gather waterbody assessment methodology information from each state, the project team developed a spreadsheet (see Appendix A) to document the following information:

- State agency responsible for developing waterbody assessments
- State agency contact information, including address, website and key personnel
- Brief description of the availability of the state's most recent Integrated Report, or 305(b) Report and 303(d) list

Additionally, the project team developed a template for compiling information (referred to herein as a "survey") to ensure a thorough and consistent evaluation of each state's documented waterbody assessment methodology. An example of the evaluation survey is included in Appendix B. The project team then obtained watershed reports (i.e., either the Integrated Report, or the 305(b) report and 303(d) list) from all 50 states, along with any associated description of assessment and listing methodology. For most states, the reports were available on their environmental department websites. The project team reviewed each report and completed the evaluation form discussed above. When the reports contained only limited documentation on assessment methodology, further searches were made to obtain sufficient detail on the waterbody assessment and listing methodology. This documentation was found in separate policy reports or guidance documents, sometimes labeled as "consolidated assessment and listing methodology," "continuing planning process" documents or other, more formalized assessment and listing policies.

2.3 Interviews with State Personnel

Completion of the literature review and evaluation survey highlighted the adequacy of the documentation of each state's waterbody assessment and listing methodology. For the majority of states, the documentation was thorough and detailed enough to understand the basic assessment process. The detailed evaluation survey was designed to obtain an in-depth understanding of each state's methodology. However, the description of methodologies in the state Integrated Reports and associated documents often was not sufficient to identify the strengths and weaknesses of each methodology. Therefore, a subset of key questions was selected from the evaluation survey and additional information was sought using telephone interviews with state agency personnel. Interviews focused on the following questions:

1. What specific types of data are required to conduct waterbody assessment methodology for listing purposes?
2. Are there specific QA/QC requirements for data used in the assessment?
3. If data do not meet specific QA/QC requirements, how are the data used in the assessment?
4. How is the assessment methodology tied to monitoring network design?

5. How does the assessment methodology extrapolate water quality over space and time where data are not available?
6. How does the assessment methodology specifically determine water quality standard attainment or non-attainment?
7. How does the assessment methodology quantify uncertainty?
8. Was public participation and/or peer review sought when developing the assessment methodology?

2.4 Development of Critical Evaluation Matrix

Information about each state's assessment methodology, gathered from the literature review and telephone interviews of state personnel, was compiled in an evaluation matrix (see Appendix C) to facilitate comparison among states. Due to the project timing, the research team modified the summaries of the states' assessment methodologies, where possible, when the states indicated that their methodology had changed from 2004. This quality control check was conducted by e-mailing the critical evaluation matrix back to each state contact person for review and comment. Any revisions suggested by the state contact person were incorporated into the final evaluation matrix shown in Appendix C.

2.5 Review of Assessment Guidance Documents

The research team gathered information on existing guidance for waterbody assessments to aid in the evaluation of state methodologies and help develop recommendations. These guidance documents included the following:

- U.S. EPA's *Guidance for 2004 Assessment, Listing and Reporting Requirements Pursuant to Sections 303(d) and 305(b) of the Clean Water Act* (U.S. EPA, 2003a)
- U.S. EPA's *Guidance for 2006 Assessment, Listing and Reporting Requirements Pursuant to Sections 303(d), 305(b) and 314 of the Clean Water Act* (U.S. EPA, 2005a)
- U.S. EPA's *Consolidated Assessment and Listing Methodology: Towards a Compendium of Best Practices* (U.S. EPA, 2002)
- U.S. EPA's series of data quality management guidance documents, including:
    - *Guidance on Systematic Planning using the Data Quality Objectives Process* U.S. EPA QA/G-4 (U.S. EPA, 2006a)
    - *Data Quality Assessment: A Reviewer's Guide* U.S. EPA QA/G-9R (U.S. EPA, 2006b)
    - *Data Quality Assessment: Statistical Methods for Practitioners* U.S. EPA QA/G-9S (U.S. EPA, 2006c)
    - *U.S. EPA Requirements for Quality Assurance Project Plans* U.S. EPA QA/R-5 (U.S. EPA, 2001)
- The Federal Water Quality Coalition's *Preparation of Integrated Water Quality Monitoring and Assessment Reports* (FWQC, 2002)
- The National Research Council's report titled *Assessing the TMDL Approach to Water Quality Management* (NRC, 2001)

- A review of U.S. EPA Region V states' monitoring and assessment programs titled "*Region V State Bioassessment and Ambient Monitoring Programs: Initial Evaluation and Review*" (Yoder, 2004)
- A joint U.S. EPA and ASIWPCA report titled "*Important Concepts and Elements of an Adequate State Watershed Monitoring and Assessment Program*" (Yoder, 1997)
- U.S. EPA's *Elements of a State Water Monitoring and Assessment Program* (U.S. EPA, 2003b)
- A report published in the *Journal of Environmental Engineering* titled "Biological Assessment and Criteria Improve Total Maximum Daily Load Decision Making" (Karr and Yoder, 2004)

2.6 Development of Recommendations

The project team compiled information regarding the states' assessment methodologies into a summary of how the states currently are performing waterbody assessments for listing purposes (described in Chapter 3.0). This summary includes the description of the most common assessment methodologies used by the states and unique methodology variations. The team then contacted the PSC and Expert Panel to obtain input regarding issues, criticisms and recommendations associated with the current methodologies employed by the states. Using feedback from the PSC and Expert Panel, along with knowledge gained from previous experience in conducting water quality data analysis and reviewing current guidance, the project team developed a number of recommendations (described in Chapter 4.0). These recommendations are intended to serve as guidance to the states during development and/or refinement of their waterbody assessment and listing methodologies. Information provided in the summary of state methodologies and recommendations also should serve as a reference tool for the states as they research ways to improve their assessment methodologies.

CHAPTER 3.0

RESEARCH FINDINGS: WHAT DO CURRENT WATERBODY ASSESSMENTS LOOK LIKE?

The following chapter provides a summary of the waterbody assessment and listing methodologies employed by the states for their most recent Integrated Report or statewide waterbody assessments (i.e., 305(b) report) and list of impaired waters (i.e., 303(d) list). A complete summary of state methodologies is included in Appendix C. A summary of U.S. EPA's current guidance regarding preparation of the states' Integrated Report provides context for the discussion of the states' assessment methodologies. The summary of assessment methodologies herein focuses on the most common methods employed by the states, with unique methodologies noted for their particular strengths in determining attainment of water quality standards with reliability and confidence.

3.1 U.S. EPA'S Integrated Report Guidance

The stated objective of U.S. EPA's Integrated Report guidance document (2005a) is to recommend a reporting format and suggested content for a single document that integrates the reporting requirements of the CWA Sections 303(d), 305(b) and 314. The purpose of each state's Integrated Report is to report on the water quality standards attainment of all waters in the state, document the availability of data and information for each water, identify certain trends in water quality conditions, and provide information to managers in setting priorities for future actions to protect and restore the health of the nation's aquatic resources. Although some states still provide separate reports for 305(b) and 303(d) purposes, 33 states used U.S. EPA's recommended Integrated Reporting format in 2004 to classify waterbodies of the state and develop the list of impaired waters (303(d) list). U.S. EPA's goal is for all states to utilize the Integrated Report format by 2008 (U.S. EPA, 2005a).

The basic structure of the Integrated Report is to provide a five-part categorization format for reporting all waters of the state. These five sorting categories are summarized in Table 3-1.

Table 3-1. U.S. EPA Integrated Report Five Reporting Categories.

Category	Description
1	All designated uses are supported, no use is threatened
2	Available data and/or information indicate that some, but not all of the designated uses are supported
3	There is insufficient available data and/or information to make a use support determination
4	Available data and/or information indicate that at least one designated use is not being supported or is threatened, but a TMDL is not needed
5	Available data and/or information indicate that at least one designated use is not being supported or is threatened, and a TMDL is needed

From: U.S. EPA (2005a)

The classification of waters into the five categories allows the states to prioritize water quality protection measures and monitoring activities, and to determine which waterbodies need a TMDL analysis (Category 5), and which waterbodies are currently being addressed by existing TMDLs or alternatives (Category 4). Using the definitions above, Category 5 becomes the state's 303(d) list of impaired waters.

In addition to the reporting categories, U.S. EPA also recommends that states use integrated monitoring and assessment techniques to both assess waterbody conditions on a statewide basis (305[b]), and comprehensively identify all impaired and threatened waters (303[d]). Specifically, U.S. EPA proposes that the states utilize a probability-based monitoring program to assess waters on a statewide scale, along with site-specific monitoring and assessments to determine individual waterbody designated use support. Specific reporting recommendations in the Integrated Report guidance (U.S. EPA, 2005a) include the proposal that states manage their assessment results in an Assessment Database (ADB) or other electronic data management system that can be submitted along with the Integrated Report.

3.2 What Specific Types of Data Are Required to Conduct Waterbody Assessment Methodology for Listing Purposes?

3.2.1 Data Sufficiency

The type of data used in the waterbody assessments is just as important to the outcome of the assessment as the data evaluation method. As required in U.S. EPA guidance documents (2003a, 2005a), most states consider "all readily available data" for waterbody assessments, provided that minimum quality and/or quantity requirements are met. With a few exceptions (Illinois, Indiana, Kentucky, Ohio), most states use not only data collected through state monitoring programs, but also data collected by federal agencies, universities, dischargers, citizen/volunteer monitoring groups, and other stakeholders.

Although states consider data from a variety of sources, states have varying requirements to ensure that the data are representative of current waterbody conditions, also termed *data sufficiency*. The data most commonly used by states to determine attainment or non-attainment of water quality standards include biological, chemical and physical (i.e., habitat) data collected through sampling. Some states consider "observational" data (qualitative data not collected through sampling) to prioritize waterbodies for future sampling efforts, but do not use them for listing purposes.

When conducting waterbody assessment for listing purposes, most states only use data collected in the most recent five years prior to the listing cycle. The minimum temporal and spatial coverage requirements of water quality samples vary widely among the states, with some states having no requirements and others requiring water quality observations to span a minimum time period and/or minimum spatial area. Texas, for example, requires data to be collected for more than one year and a minimum of two sampling stations per waterbody. As illustrated in Figure 3-1, thirty-two states require a minimum number of water quality samples in order to perform an assessment for listing purposes. Minimum sample sizes range from 3 to 24, though most states use 10 samples as a minimum data requirement to make impairment determinations.

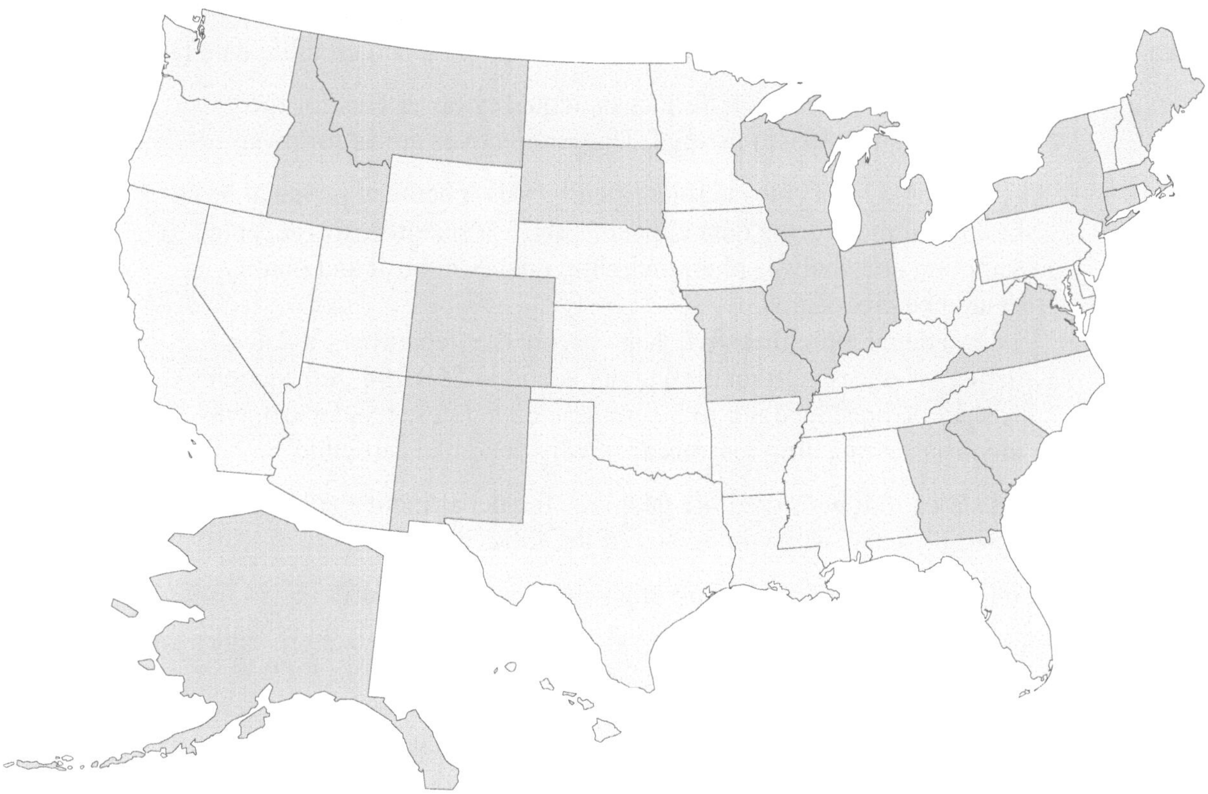

Figure 3-1. States With Minimum Sample Size Requirements (shown in the lighter color).

For all states with specific data requirements, waterbodies that do not have sufficient data are not assessed, and if using the Integrated Report format, are grouped into Integrated Report Category 3, which requires additional data collection before a determination of support or non-support of designated use can be made. Several states (e.g., Idaho, Indiana, New Hampshire, Utah, Nevada and Wyoming) do allow the caveat that overwhelming evidence or evidence of adverse effects, which varies in definition, can be used to determine impairment for waterbodies which otherwise would not meet minimum data requirements. For example, Utah requires a minimum of 10 samples to determine impairment for chronic toxic pollutants; however, the state may determine impairment if a waterbody has less than 10 samples, but one sample has a concentration greater than 1.75 times the chronic criterion.

3.2.2 Data Quality Requirements

As defined by U.S. EPA (2006d), data quality objectives are qualitative and quantitative statements that clarify objectives, define appropriate types of data, and specify levels of potential decision errors that will be used as the basis for establishing the quality and quantity of data needed to support decisions. U.S. EPA requires the states to implement data quality management programs for those monitoring and data collection activities that are conducted by the states (U.S. EPA 2001, 2006a, 2006b, 2006c). Many of the states have extended these requirements to data collected from outside data sources. Numerous states have minimum data quality requirements for data that are to be included in the waterbody assessment process.

Seven states (Arizona, Florida, Iowa, Montana, Ohio, Washington and Wyoming) have "credible data" laws which require that data meet minimum QA/QC requirements to be used for waterbody assessments, particularly for inclusion on the list of impaired waterbodies. Portions of Iowa's law are given below as an example of elements included in a credible data law:

> Iowa's credible data law defines data quality objectives for the State's Section 303(d) listings. These objectives are as follows:
>
> - "Credible Data" means scientifically valid chemical physical or biological monitoring data collected under a scientifically accepted sampling and analysis plan, including quality control and quality assurance procedures.
> - Data dated more than five years before the department's date of listing or other determination under section 455B.194, Subsection 1, shall be presumed not to be credible data unless the department identifies compelling reasons as to why the data is credible.
>
> As stated in Iowa's credible data law, the department shall use "credible data" when doing any of the following:
>
> - Developing and reviewing any water quality standard.
> - Developing any statewide water quality inventory or other water assessment report (Note that Iowa's 305(b) assessments are not subject to requirements for credible data.)
> - Determining whether any water of the state is to be placed on or removed from any Section 303(d) List.
> - Determining whether any water of the state is supporting its designated use or other classification. (Note that the credible data law does not require the use of credible data for establishment of a designated use or other classification of a water of the state.
> - Determining any degradation of a water of the state under 40 CFR 131.12.
> - Establishing a TMDL for any water of the state. (Iowa DNR, 2005)

Thirty-nine states have data QA/QC requirements for data collected outside of state monitoring programs. These QA/QC requirements are often summarized in the state's Integrated Report, though references to separate QA/QC and sampling protocol guidance documents are often cited as well. For data that do not meet specific QA/QC requirements, the states usually pursue one or more of the following options:

- The data are not used in the waterbody assessment
- The data are used only as evidence to target the waterbody for additional monitoring
- An assessment is performed with the data, but the assessment is considered "evaluated" and not "monitored" (less confidence is associated with an evaluated assessment)
- If no other data exist for the waterbody, it is placed in Integrated Report Category 3

3.3 How Is the Assessment Methodology Tied to Monitoring Network Design?

The states have recognized the importance of monitoring design in providing scientifically defensible data for assessment and decision-making purposes and have implemented monitoring programs at various scales. Because states do not have the resources to monitor all waters of their state, they often must rely on data collected through monitoring programs for which the monitoring purposes were not the same as the information objectives of the state. Recognizing this, a few states (e.g., Minnesota, Idaho, Indiana, Delaware, Florida, and New Hampshire) rely mainly on data collected through state monitoring programs designed for their statewide water quality assessment (305(b) report), which often are obtained through a probability-based monitoring network. The design of these probability-based monitoring networks allows the states to sample a subset of all waters and then provide a statistically derived estimate of the quality of all waters along with a statement of the uncertainty of that estimate. U.S. EPA's Environmental Monitoring and Assessment Program (EMAP) (www.epa.gov/EMAP) provides more information on probability-based sampling.

Although states do perform monitoring to support waterbody-specific assessments to determine water quality standards attainment for 303(d) list development, no state uses an monitoring network designed exclusively for this purpose. Reasons for this include the following:

- Limited monitoring resources
- Multiple data needs of the state (i.e., not only for Integrated Report assessments, but also for water quality standards setting, National Pollutant Discharge Elimination System [NPDES] compliance, etc.)
- The U.S. EPA requirement that states consider all readily available data in the waterbody assessment process (U.S. EPA, 2005a)

Therefore, it is difficult to anticipate the type and quality of data to be used in future waterbody assessments, so the states must allocate resources to determining the sufficiency and credibility of each data source during every assessment and listing cycle.

For most states the monitoring network is not specifically designed to support their water quality assessment methodology. Rather, it functions to gather data for multiple uses.

3.4 How Does Assessment Methodology Extrapolate Water Quality Over Space/Time Where Data Are Not Available?

As discussed above, states do not have the resources to monitor all waters of the state and so must rely on outside data sources for waterbody-specific assessments. Often, many waterbodies will have limited or no data available. In order to address this issue, 12 states have segmented their waterbodies into assessment units (AUs), reaches or segments for which the outcome of a single assessment can be applied to all waters within that AU. These AUs can represent a single lake or reservoir, or several miles of a stream or river, and are often developed to represent waters with assumed homogenous water quality.

The development of AUs differs among the states, though all use some type of watershed-based segmentation, most often tied to a spatial database of the watersheds of their state. Some states rely on the USGS Hydrologic Unit Code (HUC) watershed designations, often at varying scales, while others rely on the National Hydrography Dataset (NHD) spatial database of watershed designations. Some states require that an assessment unit have at least one sampling station, using the locations of sampling stations to aid in the AU segmentation. Additional factors used to segment waters include changes in the waterbody condition, such as the introduction of a point source discharge, a tributary, or a land use change. For those states which do not extrapolate assessment results using AUs, four states (Oklahoma, Nebraska, Illinois, and Texas) specifically limit the length of stream for which an assessment result can apply, typically no more than 25 miles upstream or downstream of the sampling station which follows U.S. EPA guidance (2003a).

As discussed in Section 3.3, for statewide waterbody assessments for 305(b) purposes, several states extrapolate water quality findings to unmonitored locations through the use of a probability-based monitoring network. However, this special extrapolation is limited to statewide water quality characterization and is not used for 303(d) listing purposes. Eleven states currently have probability-based monitoring networks (Maryland, Washington, Minnesota, Oregon, Idaho, Indiana, Connecticut, Florida, Kentucky, Maine, and New Hampshire).

Temporal data extrapolation is assumed to occur because many states are taking grab samples and comparing them to chronic criteria, though the process is not well-documented by the states. After review of the states' assessment methodologies, it is not clear how states extrapolate grab-sample data to evaluate compliance with water quality criteria with specific durations (i.e., 1-hour, 4-day, 30-day averages).

For listing purposes it appears that states are not extrapolating data beyond monitored waters for impairment determination.

3.5 How Does Assessment Methodology Specifically Determine Water Quality Standard Attainment or Non-Attainment?

U.S. EPA (2006d) has defined water quality standards in the following manner:

> Water quality standards serve as the foundation of the water quality-based control program mandated by the Clean Water Act. Water quality standards define the goals for a waterbody by designating its uses, setting criteria to protect those uses, and establishing provisions to protect water quality from pollutants. A water quality standard consists of four basic elements:
>
> (1) ***designated uses*** of the waterbody (e.g., recreation, water supply, aquatic life, agriculture),

(2) ***water quality criteria*** to protect designated uses (numeric pollutant concentrations and narrative requirements),

(3) an ***antidegradation policy*** to maintain and protect existing uses and high quality waters, and

(4) ***general policies*** addressing implementation issues (e.g., low flows, variances, mixing zones).

States, in accordance with this definition and regulatory requirements, have developed water quality standards to protect designated uses for waterbodies of the state. These designated uses most often are determined as the attainable beneficial use for the specific waterbody, which may or may not be the current existing use. In order to determine the water quality condition that will support the designated use, states have established numeric criteria which serve as thresholds by which physical, chemical, and biological indicators are compared to determine support of designated uses and thus attainment of water quality standards. In many states, however, numeric criteria have not been established for all water quality standards.

Because water quality criteria and standards are designated use-driven, states have developed evaluative methods which are specific to the designated use (e.g., aquatic life, drinking water, recreation, fish consumption) and type of water quality indicator (e.g., physical, chemical or biological). For example, the method to determine aquatic life use support may utilize a bioassessment approach that examines the existing aquatic community and compares it to a reference condition, while the methodology to determine recreation use support may look only at strict compliance with numeric bacteria criteria.

For waterbodies that are not attaining water quality standards, states are encouraged to determine the source and cause of impairment and determine if the waterbody could attain water quality standards through the development of a TMDL. Those waterbodies are included in the Integrated Report Category 5, also known as the 303(d) list of impaired waters. The majority of states make the determination that if a waterbody is not exceeding water quality criteria for which data are available, then the waterbody is attaining its designated use and meeting water quality standards. However, a few states (e.g., Oklahoma, Arizona, New Hampshire, New Jersey, West Virginia) have established a set of core indicators, for which water quality observations must meet water quality criteria before that waterbody can be determined as fully supporting its designated use. Those waterbodies are assigned to Integrated Report Category 1. An example of core indicators used by the state of Arizona is provided in Table 3-2.

Table 3-2. Core Indicators Used by the State of Arizona to Determine Attainment Status of Waters.

CORE PARAMETRIC COVERAGE For each designated use, at least three samples of the following parameters are required to assess the designated use as "attaining" uses:	
Designated Use	**Parameters**
Aquatic and Wildlife	Dissolved oxygen, flow (if a stream) and depth (if a lake), hardness, pH, turbidity/suspended sediment concentration, total nitrogen and total phosphorus[1], dissolved metals (cadmium, copper, zinc)
Fish Consumption	total mercury
Full Body or Partial Body Contact	*Escherichia coli*, pH
Domestic Water Source	nitrate/nitrite or nitrate, pH, total fluoride, and total metals (arsenic, chromium or chromium VI and lead)
Agricultural Irrigation	pH, total boron, and total manganese
Agricultural Livestock Water	pH, total copper, and total lead

From: Arizona Department of Environmental Quality (ADEQ) (2004)

Special Notes:

1. Nitrogen and phosphorus are required only in surface water with nutrient standards.
2. Dissolved oxygen, turbidity/suspended sediment concentration (SSC) and *Escherichia coli* are not required in ephemeral waters.
3. SSC is not required in effluent dependent waters.

For non-toxic (i.e., conventional) chemical pollutants with established numeric criteria, such as dissolved oxygen or pH, most states (32) utilize the "ten-percent rule" to determine compliance (see Figure 3-2). The ten-percent rule is based on guidance from U.S. EPA (1997), which recommends making a determination of non-attainment when more than 10% of measurements exceed the water quality criteria for conventional pollutants.

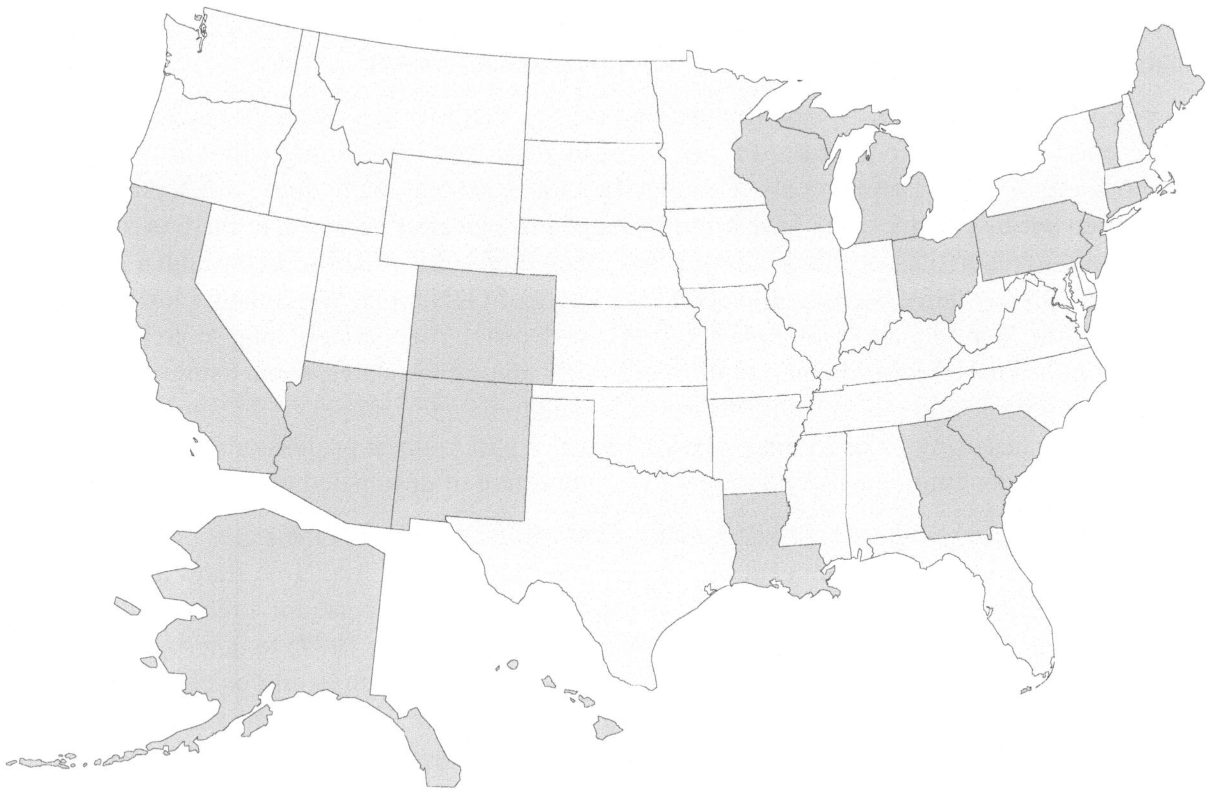

Figure 3-2. States Using the Ten-Percent Rule to Determine Compliance with Specific Water Quality Standards (shown in the lighter color).

Both statistical and non-statistical methods are used to determine if ten percent of water quality samples from a waterbody exceed water quality criteria. The majority of states (25) use a non-statistical procedure in which a simple percentage of samples exceeding the water quality criteria is calculated (see Smith et al., 2001 for discussion of the "raw-score" method). Several states (California, Washington, Kansas, Nebraska, Alabama, Arizona, Florida and Texas) use a statistically based methodology known as the *binomial method*, which estimates the probability of the true excursion rate of the population being above the acceptable excursion rate (e.g., 10 percent), given the number of water quality observations or samples. Other states (Colorado, Delaware) compare upper percentiles of the data to the numeric criteria to determine attainment. For example, Colorado uses a raw-score estimate of the upper 85th percentile, while Delaware uses a non-parametric statistical estimate of the upper 90th percentile, both of which recognize variability in environmental data. Using a statistically based methodology allows the state to determine uncertainty associated with the assessment, which is helpful in determining attainment of water quality standards and in TMDL priority-setting purposes.

While some states continue to apply the ten-percent rule to determinations of impairment due to toxic chemicals, many state assessments of toxic chemicals for determining use support of aquatic life do not use the ten-percent rule. Rather, 18 states have numeric criteria established based on the requirement that the criteria—either acute (one-day) or chronic (four-day or thirty-day)—not be exceeded more than once in three years. For these indicators the evaluative method conducted by the states follows the same requirement, that is, more than one exceedance of toxics in a three year period results in a determination of impairment. This methodology, while contested by some, provides a clear connection between the water quality criteria, which addresses both magnitude and frequency, and the assessment methodology used to determine impairment. Such a connection is not present for most other types of criteria.

For determination of aquatic life use support, the majority of states (36) use biological assessments, either alone or as part of a weight-of-evidence approach along with chemical indicators and/or habitat assessments. The specific bioassessment methodology differs among states, often because many states have not developed biological or habitat criteria. (As of this review, biocriteria or site-specific biological reference thresholds established through a public rule-making process have been established in the states of Oklahoma, Wisconsin, Florida, North Carolina, Ohio, Vermont and Maine). Therefore, the decision rules establishing reference threshold values for biological community scores vary among the states, some using an ecoregion approach, and others using site-specific data to develop reference threshold scores. Many of the states using bioassessments rely, at least in part, on best professional judgment to determine if resident biological communities are supportive of designated uses.

Nutrients are another set of water quality indicators for which some states perform assessments without the establishment of numeric water quality criteria. Nine states assess nutrients as part of determining support of recreation and aquatic life use for streams, lakes and reservoirs. Four of these states use the Trophic State Index (Carlson, 1977) to determine the trophic status of each waterbody. For those lakes and reservoirs which would be classified as eutrophic, a determination of impairment is made, although there is not a clear connection made between existence of eutrophic conditions and impairment of designated use.

Eleven states (see Figure 3-3) use multiple indicators (i.e., data types) to determine use support, and portray this process in their Integrated Reports as a "weight-of-evidence" approach. This type of approach supplements chemical data with other data types—including biological, habitat, and toxicity tests—to assess support of aquatic life uses, human health, fish consumption, wildlife, agriculture and other measures. Many states consider multiple indicators to reduce uncertainty in determining use support; however, the final determination of waterbody impairment often is based on only one data type indicating impairment. This is because of the longstanding U.S. EPA policy of independent applicability, which supports the approach that if one line of evidence points to impairment, then the waterbody is considered impaired. Because U.S. EPA has final review and approval of the state's 303(d) list of impaired waters, the function of the independent applicability principle has important consequences to the state's assessment methodology development and allocation of resources for both monitoring and data evaluation and assessment, particularly in the use of bioassessments along with chemical data to determine impairment/non-impairment.

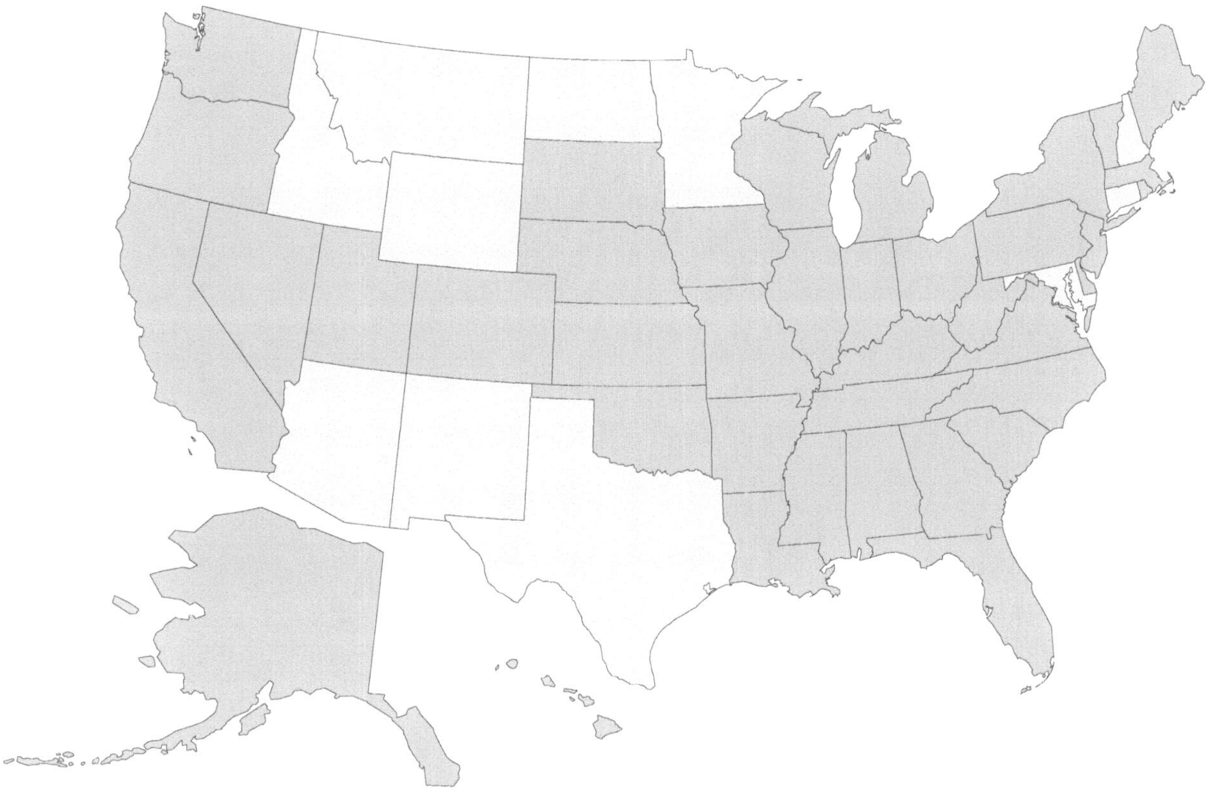

Figure 3-3. States Which Use Weight-of-Evidence Approach to Determine Use Support (shown in the lighter color).

In sum, state assessment methodologies determine water quality standard attainment or non-attainment in a variety of ways, depending on the nature of each state's available monitoring data and the designated use or uses of the waterbody.

3.6 Does Assessment Methodology Quantify Uncertainty?

Few states try to quantify the uncertainty associated with their waterbody assessments. Thirteen states (see Figure 3-4) use methodologies in which uncertainty can be calculated or inferred based on the use of statistically based analysis methods with a reported confidence level. Although this quantification of uncertainty addresses only the uncertainty associated with the particular data analysis method, it at least provides a way to communicate the confidence associated with the determination of impairment/non-impairment to the public and other stakeholders.

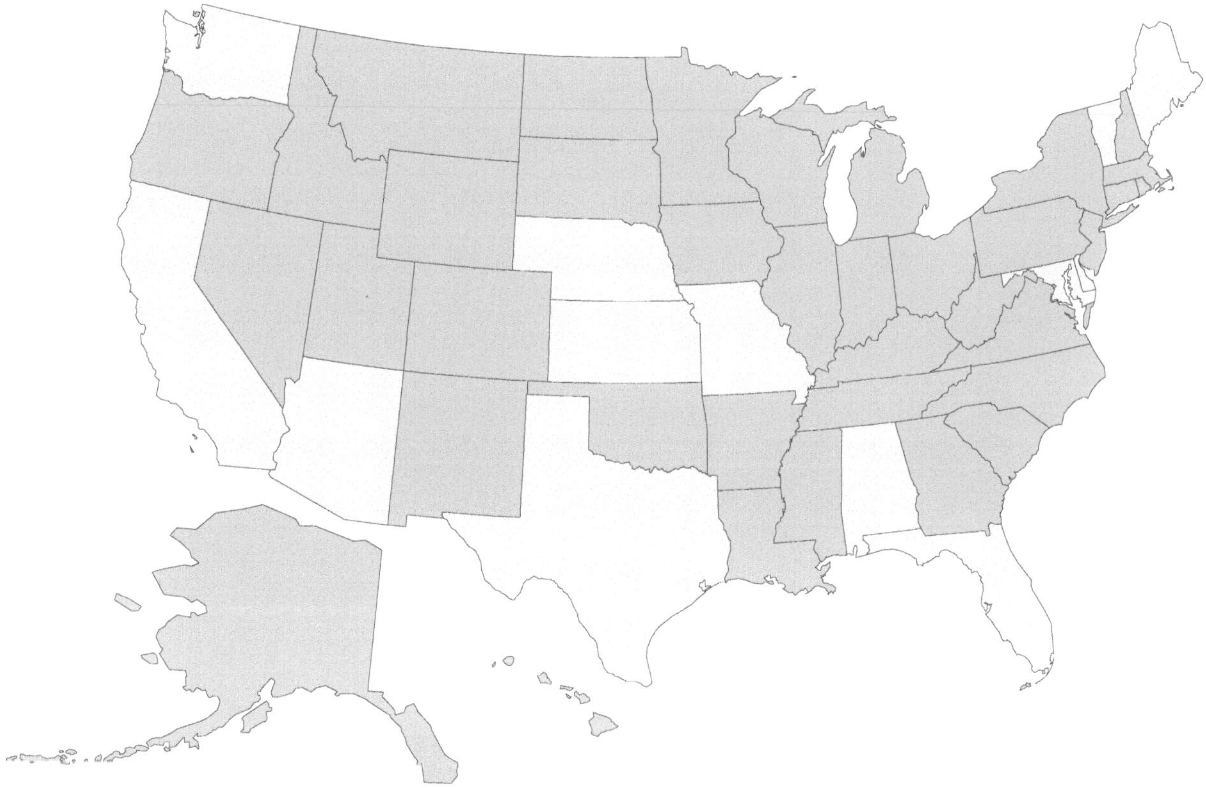

Figure 3-4. States Which Quantify Uncertainty in Their Assessments (shown in the lighter color).

Although state waterbody assessments have inherent uncertainty associated with them, many states claim the need to be precautionary in their determination of use support. In other words, they believe it is better to err on the side of non-attainment of designated use than to err on the side of attainment. This policy decision, which assumes no negative consequences when resources are unlimited, very often is a driving factor in the selection of a state's waterbody assessment methodologies. Assuming that the states start with a presumption that the waterbody is meeting water quality standards (e.g., null hypothesis), the result of this precautionary approach is that the states' waterbody assessment methodologies tend to minimize the probability of Type I error (false positive or false non-attainment) rather than Type II error (false negative or false attainment).

3.7 Was Public Participation/Peer Review Sought When Developing Assessment Methodology?

Though not required for 305(b) assessments, U.S. EPA requires that the 303(d) list of impaired waters be published for public review and comment. U.S. EPA Integrated Report guidance (2005a) also recommends that the states make their assessment methodologies available to the public for review and comment. It was found that most states include the public by publishing their draft 303(d) list for review and comment. Often this list includes a summary of the assessment methodology.

Many states solicit data for waterbody assessments from the public and/or stakeholders through newspaper notices, e-mail and telephone contact. However, the majority of states do not allow public participation in the methodology development prior to performing the waterbody assessments. Only 18 states publish their draft assessment methodologies for public review and comment prior to development of the draft list of impaired waters. Other states submit their methodology for public review along with the draft list of impaired waters. Ohio and Utah also use a peer-review process during methodology development.

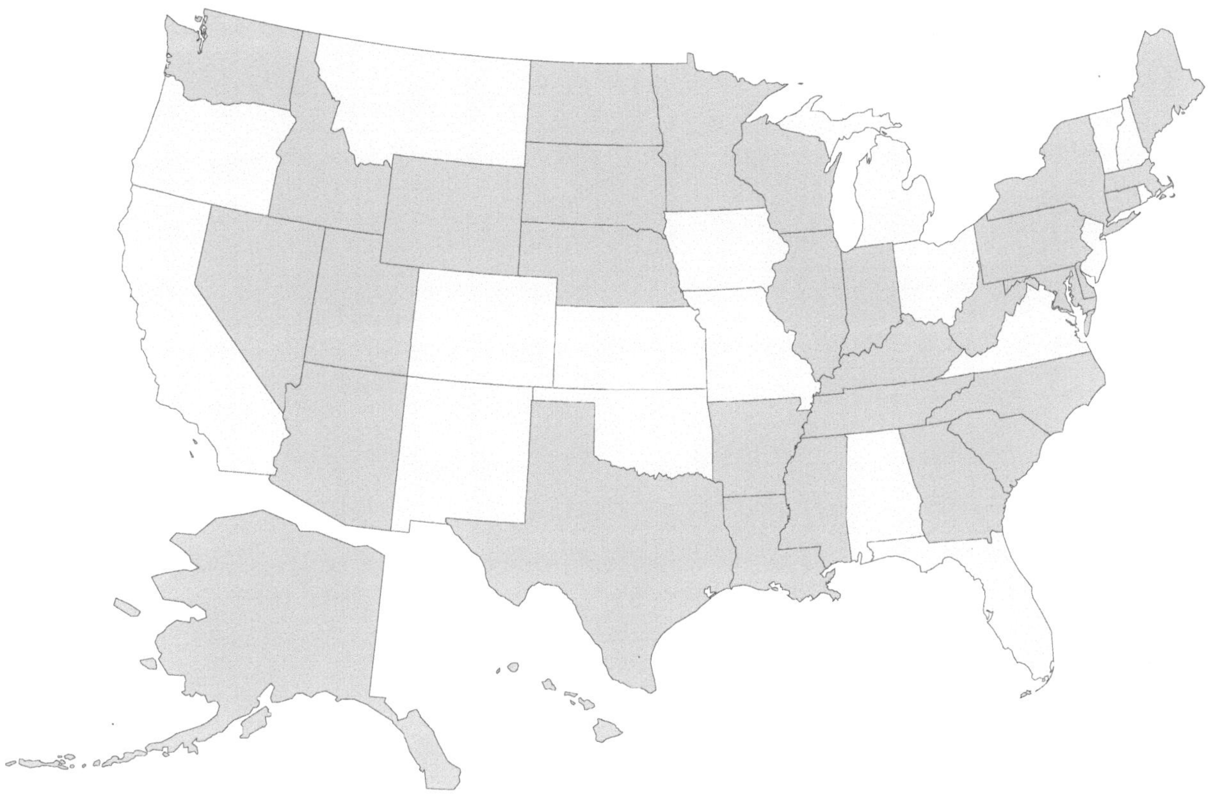

Figure 3-5. States Which Solicit Public Comment During Assessment Methodology Development (shown in the lighter color).

CHAPTER 4.0

RECOMMENDATIONS

Chapter 3.0 provides a summary of the current waterbody assessment and listing methodologies and how states optimize available data and characterize waterbody conditions and uncertainty. This summary was developed by compiling information regarding the states' current assessment methodologies under several key evaluation questions. This chapter takes those same key evaluation categories and provides additional discussion and recommendations regarding how to improve reliability of the waterbody assessment and listing process.

Although each state needs to tailor its waterbody assessment process to its regulatory requirements and water quality protection programs, consistency among the states is desirable from a national perspective. In reviewing the Integrated Reports and assessment methodologies for all 50 states, it became apparent that the states are in various stages of developing their waterbody assessment programs. Some states have addressed much of the uncertainty inherent in performing these assessments by using scientifically defensible data and analysis methods and making their waterbody assessment process transparent and accessible to the public. Other states are still working to develop methodologies which will allow them to more confidently determine attainment or non-attainment of water quality standards. The purpose of this chapter is not to provide a prescriptive waterbody assessment process to be followed by all the states, but rather to provide recommendations that will serve as a reference tool and guidance to the states who are working to improve their waterbody assessment methodologies.

The NRC (2001) notes that a formalized process is needed through which states can capitalize on lessons learned, transfer technology or share knowledge. The research team recommends that U.S. EPA develop such a process for dialog among the states, perhaps using the findings from this report as a review of current status and an indication of where improvement is needed. Another forum for technology transfer could be the National Water Quality Monitoring Council, which facilitates a biannual water quality monitoring and assessment conference well-attended by state and federal agency employees.

Many of the following recommendations have been included in recent Integrated Report guidance (U.S. EPA, 2003a, 2005a). These recommendations are reiterated here because the literature review of state assessment methodologies indicates that progress is still needed to comply with these recommendations.

4.1 Data Requirements to Conduct Waterbody Assessment Methods for Listing Purposes

4.1.1 Develop and Publish Minimum Data Quality Requirements

Many states have established stringent data requirements that must be met in order to incorporate data into the waterbody assessments for listing purposes; however, not all states have formalized this process. This is recognized by several members of the water quality community, including the Federal Water Quality Coalition (FWQC). "State water quality monitoring and assessment programs should establish quality assurance project plans (QAPPs) and refine data quality objectives to improve the process of collecting data and information for regulatory and planning purposes" (FWQC, 2002).

States are required to assemble and evaluate "all existing and readily available water quality-related data and information" to develop the 303(d) list, but verifying the quality of data from outside sources may be difficult. For those states that have not formalized minimum data quality requirements for outside data sources, the research team recommends this as an important first step to improving the consistency and reliability of assessments to determine use support. It also is imperative that these data requirements be published and disseminated to the public so that stakeholders who provide monitoring data to the states can incorporate the requirements into their monitoring programs and data management systems. For outside entities that continually provide data to the waterbody assessment process, states should request that those entities submit a QAPP for review and approval. An approved QAPP will enhance confidence in the data provided by the entity. The benefits anticipated to result are summarized in Figure 4-1.

Figure 4-1. Benefits of Developing and Publishing Minimum Data Requirements.

Data requirements should include a statement of monitoring objectives, minimum accompanying metadata (e.g., sampling location, sampling date, analytical method), minimum spatial and temporal coverage, minimum sample sizes, and documentation of data verification (QA/QC) procedures. States also should document which datasets do not meet minimum data requirements and are not used in waterbody assessments, because U.S. EPA asserts that it will review excluded data and the rationale for exclusion as part of the review of the state's 303(d) list (2005a). For further guidance on the appropriate use and documentation of third-party data in the state's waterbody assessment process, the states should review *A Guide to Developing Secondary Information Products: Methods Review and Documentation* (U.S. EPA, 2003c).

4.1.2 Describe How Waterbodies Will Be Assessed If Data Quality Requirements Are Not Met

Often it is not clear in the description of states' waterbody assessment methodologies how waterbodies are assessed if the data are of insufficient quantity or quality. Many states declare that such waterbodies will be placed into Integrated Report Category 3 (unassessed) and do not consider additional evaluation of that waterbody until more data are made available. However, some states will disregard data quality requirements, such as minimum sample size, if "overwhelming evidence" is available to justify a determination of impairment. Few states provide a definition of overwhelming evidence, or the evaluative method that would be used to determine if such evidence is sufficient to determine impairment.

The research team recommends that if a state chooses to allow "overwhelming evidence" as an exception to its minimum data requirements, a specific methodology for verification of the data be developed and published, along with decision rules for defining overwhelming evidence and an explanation of how it will be used to assess waterbodies for use attainment.

4.1.3 Address Data Censorship Issues in Assessment Methodology

In their discussions of QA/QC requirements, very few states address the issue of data censorship—such as removing or replacing non-detect values or removing outliers—prior to assessment. The presence of both non-detects and outliers can impact the outcome of the assessment, particularly for those assessments that use the raw-score method to determine compliance with standards, those that calculate percentiles, or those that use parametric tests for trending. The methodology developed for data validation review and data censorship should be included in the documentation of the waterbody assessment methodology.

Non-detects, or low-level values below the lab's reporting level, often are found in environmental datasets and, depending on how they are handled, can introduce errors into estimations of central tendency, variability and the outcome of statistical tests. Helsel (2005a) highlights the weaknesses of the three main approaches used by environmental data analysts to handle non-detects: (1) substitution using a fraction (usually one-half) of the reporting limit, (2) use of the "machine readings" rather than allowing the lab to use a reporting limit, and (3) substitution with the reporting limit value. Recent U.S. EPA guidance (2004) also has recognized the fallibility of substitution methods, and does not recommend them; therefore, the use of substitution methods is not considered appropriate for data analysis used to determine water quality impairment. However, guidance still recommends older methods such as Cohen's method, which is valid only for data with one detection limit, and Aitchison's method, which essentially is substitution (usually but not always with 0) by another name.

Instead of substitution, Helsel (2005a) recommends three more accurate methods for computing statistics on data with non-detects: (1) maximum likelihood estimation (MLE), (2) imputation, and (3) the Kaplan-Meier method. Details of computing these and other methods for data with non-detects, along with examples, are found in Helsel (2005b). Each of these methods is available in the survival analysis or reliability analysis sections of commercial statistics software such as Minitab or SYSTAT. In particular, the Kaplan-Meier method is easy (just a counting procedure), non-parametric, and is the standard method for determining percentiles in medical and industrial statistics.

Assessing outliers in the dataset is important, particularly for those states that allow overwhelming evidence to justify determination of impairment. These "extreme" data points may not be relevant for assessing water quality for those constituents for which water quality criteria are intended to represent long-term average conditions. Data collected during extreme or rare events could skew the evaluation of central tendency, particularly if the water quality criteria are not applicable to such events. In order to ensure that data which has been deemed relevant to the assessment but which has been collected during rare or extreme events do not disproportionately affect the outcome of the waterbody assessment, weighted measures should be used. The weights should represent the proportion of time and/or space represented by each value. Data values representing a small fraction of the time or area should have proportionally lower weight in the resulting outcome.

4.2 How the Assessment Method Is Tied To Monitoring Network Design

4.2.1 Integrate Statewide Assessment Monitoring with Waterbody-Specific Assessment Efforts

State monitoring programs vary greatly in purpose, design and scale. Some states have chosen to allocate monitoring resources in a way which allows for a statewide assessment of watershed health based on a probability-based monitoring network design. Data generated from these monitoring networks are used primarily for 305(b) assessment purposes. Other states have allocated resources to allow for short-term intensive monitoring of specific watersheds, often on a rotating-basin schedule. These data often are used for many purposes, including both 305(b) and 303(d) assessments, National Pollutant Discharge Elimination System (NPDES) compliance, and water quality standards development. Some states perform monitoring designed for 305(b) and 303(d) assessments on a watershed scale, but only for biological data.

U.S. EPA has recommended ten elements that should be addressed by a state monitoring and assessment program (U.S. EPA, 2003b):

- Monitoring Program Strategy
- Monitoring Objectives
- Monitoring Design
- Core and Supplemental Indicators
- Quality Assurance
- Data Management
- Data Analysis/Assessment
- Reporting

- Programmatic Evaluation
- General Support and Infrastructure Planning

Development, documentation and implementation of these elements is expected for all states by the year 2014. The monitoring program strategy element for all states should include a clear connection and integration of monitoring conducted for 305(b) and 303(d) purposes. The monitoring strategy should require monitoring data to be collected in such a way that evaluation of attainment of water quality standards is a guiding objective, either on a statewide scale, a site-specific scale, or both. A comprehensive and integrated approach to water quality monitoring and assessment would not only streamline the waterbody assessment process, but allow the waterbody assessment program to integrate with other water quality programs. Such an approach would improve overall efficiency and aid in resolving disparate goals (Yoder, 2004).

4.2.2 Allocate State Resources to Long-Term Monitoring on Watershed and Site-Specific Scales

Many states have implemented monitoring programs aimed to better understand small-scale watershed status. These programs, often occurring on a rotating-basin scale, provide data not only for waterbody assessments for 303(d) purposes, but also to support water quality standard development, NPDES programs and TMDLs. However, national review of state water quality monitoring efforts has found that inadequate data are being collected for meaningful data interpretation (GAO, 2000), thus the states' reliance on outside data sources. Karr and Yoder (2004) state that:

> Historic neglect of the strategic aspects of monitoring has unintended consequences. First, lack of quality data limits our ability to accurately define impaired waters causing erroneous listings and failure to list impaired water bodies. Second, data adequate to diagnose causes of impairment, a crucial step in the development and implementation of TMDLs, are not available. Third, the lack of commitment to monitoring seriously impedes refinement of TMDL programs.

Guidance regarding the development of long-term, systematic monitoring and assessment programs that can provide data to fulfill information needs of all of the state's water quality programs can be found in Yoder (1997, 2004).

It is important that states allocate resources to monitoring at spatial scales appropriate for waterbody-specific assessments. Karr and Yoder (2004) warn that "States that do not monitor watersheds at thoughtfully defined geographic scales risk missing the opportunity to better understand the effects of specific human actions." This monitoring should include not only chemical measures, but physical and biological as well. When chemical and biological data are combined, fewer chemical data are required to demonstrate significant progress toward attainment of designated uses. Karr and Yoder (2004) further note that "systematic standardized monitoring strengthens our ability to connect stressor, exposure and response gradients, as well as reduces both Type I (false positive) and Type II (false negative) assessment errors."

Yoder (2004) found that unpredictable support for long-term, routine monitoring undermines three important functions of monitoring and assessment: (1) building a systematic and consistent database to support development of new and improved [assessment] tools and criteria, (2) addressing emergency needs as they arise, and (3) developing trends through time. Karr and Yoder (2004) recommend a minimum of 15-20% of state water program resources be dedicated to monitoring and assessment. Training and orientation for program managers should emphasize the need for and utility of improved monitoring and assessment, including case studies of how long-term monitoring databases have supported development of new and improved tools and criteria and identification of long-term trends (Yoder, 2004).

Figure 4-2 summarizes the benefit of allocating state resources specifically to provide long-term monitoring on both watershed and site-specific scales.

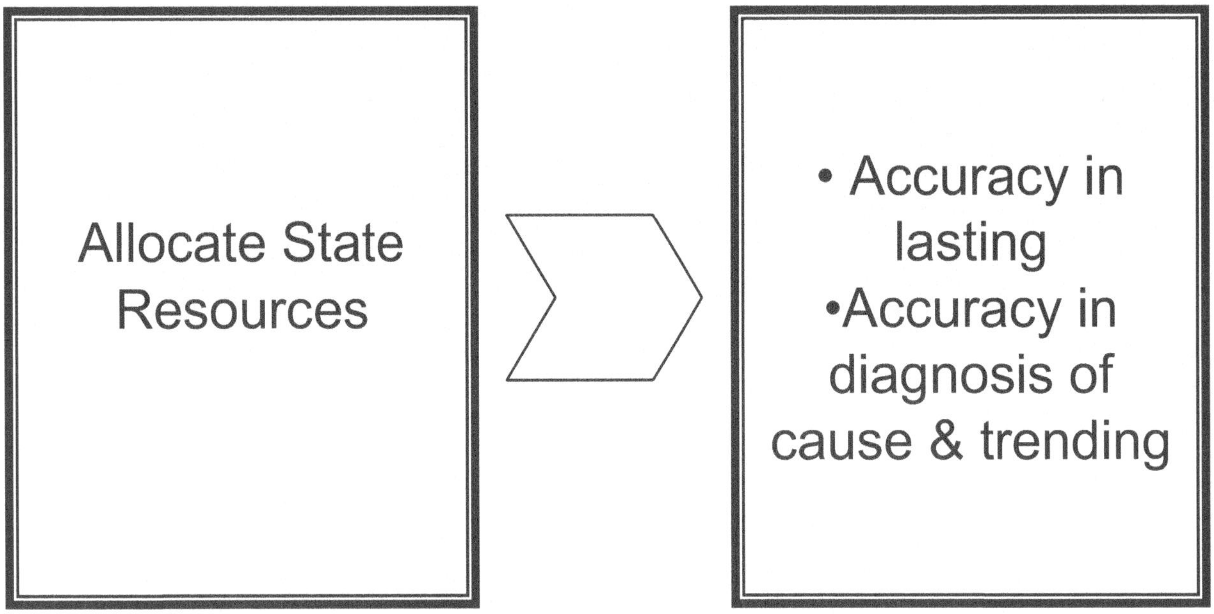

Figure 4-2. Benefits of Allocating State Resources to Long-Term Monitoring at Watershed and Site-Specific Scales.

4.2.3 Review Monitoring Objectives for Outside Data Sources

Soliciting and using data from outside data sources for waterbody assessment purposes, particularly "targeted" monitoring data, runs the risk of highlighting water quality that is only present under specific waterbody conditions. For example, the state may call for estimates of average conditions whereas the outside data may represent unusual conditions such as storms or drought. States should require a disclosure of the monitoring objectives for the monitoring programs which produce data used by the states in waterbody assessments. Review of these monitoring objectives will allow the states to properly integrate data from outside sources with data collected by the states. For example, lower weights may be applied to outside data for statistical analysis purposes if the quality of that data is uncertain. In this way, data from both random sampling and from targeted sampling can be combined for purposes of determining attainment or non-attainment of water quality standards, providing the greatest benefit for the monitoring dollar.

4.2.4 Develop Biological Monitoring and Assessment Program to Support Biocriteria Development

Using biological assessments to determine waterbody impairment is an important aspect of many state monitoring and assessment programs. However, most states do not have a transparent process for determining reference conditions by which to measure use attainment. Development of site-specific biocriteria would allow these states to more reliably and consistently determine use support. Integral to the development of biocriteria is a rigorous biological monitoring and assessment program. It is recommended that states which rely on biological data for waterbody assessment purposes focus efforts on monitoring for biocriteria development, which for some states may require additional funding mechanisms and providing incentives to stakeholders to aid in data collection efforts. Included in this effort should be a plan to include monitoring for potential stressors, as a finding of impairment must be linked to a cause (e.g., pollutant) for listing and TMDL purposes. Figure 4-3 summarizes the benefits of developing a biological monitoring and assessment program to support biocriteria development.

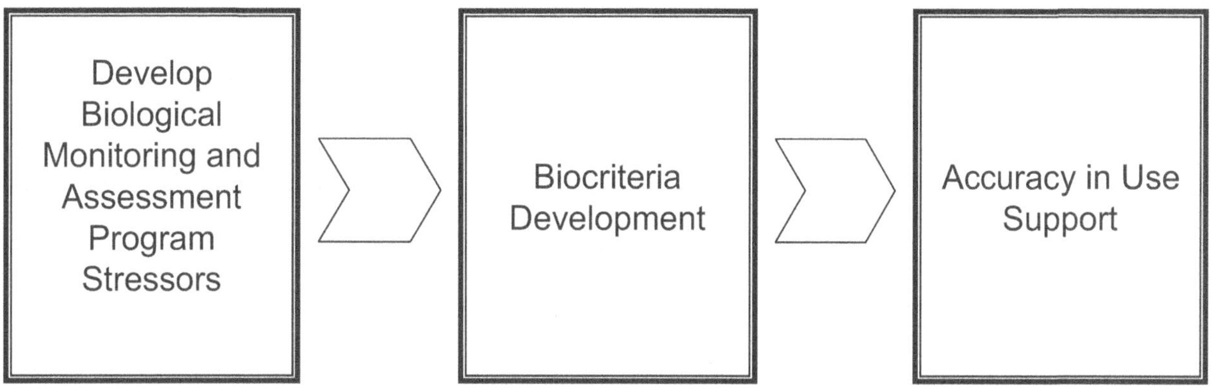

Figure 4-3. Benefits of Developing Biological Monitoring and Assessment Program to Support Biocriteria Development.

Because of the variety of methods that states use to monitor and assess water quality using biological data, Yoder (2004) recommends that a research study be conducted to evaluate the influence of biological assessments on the listing process.

4.2.5 Consider Emerging Evaluative Techniques to Analyze Data from Multiple Sources

As described in Section 3.5, the states historically have relied on two data evaluation techniques to determine attainment of numeric water quality criteria for conventional pollutants, the raw-score method and the binomial method. Both of these methods are described and evaluated in Smith et al. (2001) in terms of their Type I (false positive) and Type II (false negative) error rates when determining water quality impairment, along with a third method based on Bayes' Theorem.

Smith et al. (2001) found that the raw-score method results in a higher likelihood of a Type I error as compared to the binomial method. They also found that the binomial method is more likely to result in Type II error than the raw-score method when using small sample sizes. (Unless otherwise noted, statements in this report addressing Type I and Type II error are based on the null hypothesis that the waterbody is meeting water quality standards.)

The third approach studied, the Bayesian method, requires the development of a probability of impairment prior to the analysis, based on either empirical data or an assumption made by the analyst, which is updated to a posterior probability of impairment using the given data. Depending on the prior probability of impairment (prior), the importance assigned to the prior in the analysis, and the selected cut-off (similar to the significance level chosen by the analyst in the binomial method), the Bayesian method can control (i.e., reduce) both Type I and Type II error as compared to both the raw-score and binomial methods.

Smith et al. (2001) admit that although results from the Binomial and Bayesian methods may be similar, the Bayesian method allows managers to set prior belief about how likely sites are to be impacted. Using information from previous waterbody assessments and surrounding sites may allow for an objective formulation of priors and make the Bayesian approach a sound alternative for waterbody assessments.

The Bayesian Approach is being considered for future waterbody assessment purposes by several entities in U.S. EPA Region 7, including the states of Iowa and Kansas and the Winnebago Tribe in Nebraska. An emerging evaluative technique based on Bayes' Theorem is currently being refined by a U.S. EPA Region 7 scientist for waterbody assessment purposes (personal communication with Don Miller, April-May 2006). This evaluative technique, termed the 2-Parameter Bayesian Approach, determines the probability of impairment using the available data and user-defined "interpreters" for each specific water quality indicator. Instead of operating on the pass-fail paradigm (i.e., the water quality data are either below or above a specific numeric criterion), the Bayesian Approach uses a linear interpreter which defines, *a priori*, the probability of impairment for a linear range (or gradient) of water quality indicator values. The Bayesian Approach addresses uncertainty associated with using data of varying quality from different sources by integrating a factor called "representativeness" into its estimation of the probability of impairment.

While still an untested methodology, the 2-Parameter Bayesian Approach, as well as other emerging evaluative techniques based on Bayes' Theorem, may allow states to address sources of uncertainty in the waterbody assessment process and better manage both Type I and Type II error in their assessments. For analysts uncomfortable with methods which do not follow "classical" statistics, another evaluative technique that is more common in the drug-testing industry but has been applied to the water quality field (McBride et al., 1993; McBride 1998, 1999) is equivalence testing. This technique allows the analyst to determine, for example, if the difference in means of two sample populations is greater than some prescribed interval, rather than being exactly equal. "An advantage of this test is that the analyst must state differences of practical significance to management, also the failure to reject the null no longer induces complacency" (McBride et al., 1993).

The research team recommends that the states and U.S. EPA closely monitor the progress of these emerging evaluative techniques for waterbody assessment purposes, facilitate a scientific peer review of these approaches and encourage their wider use where appropriate.

4.3 Water Quality Extrapolation Over Space/Time Where Data Are Not Available

4.3.1 Geo-Reference Water Quality Monitoring Data

In order to ensure that monitoring data used in the waterbody assessments provide an accurate spatial interpretation of water quality, it is important that the data be geo-referenced. This will allow sampling locations to be mapped using x,y-coordinates in a geographic information system (GIS) and guarantee linkage between the monitoring data and assessment unit for which the data are intended to represent.

4.3.2 Flow-Reference Water Quality Data

Along with locational attributes, data also should be flow-referenced so that pollutant concentrations can be compared and so that trends and data extrapolation can consider appropriate flow information. For example, because concentrations change with flow, a concentration taken at a certain point in time and certain flow may not exceed the water quality criteria, however, at a different point in time and different flow, the criteria may be exceeded at the same station. Incorporation of flow consideration into water quality data interpretation is discussed by Helsel and Hirsch (2002). Better integration of water quality and water flow monitoring by the states could improve understanding of flow effects on designated use attainment (SWAQ, 2006). The requirement for flow-referenced data should extend to data received from third-party sources.

Figure 4-4 summarizes the benefits of using flow-reference water quality data.

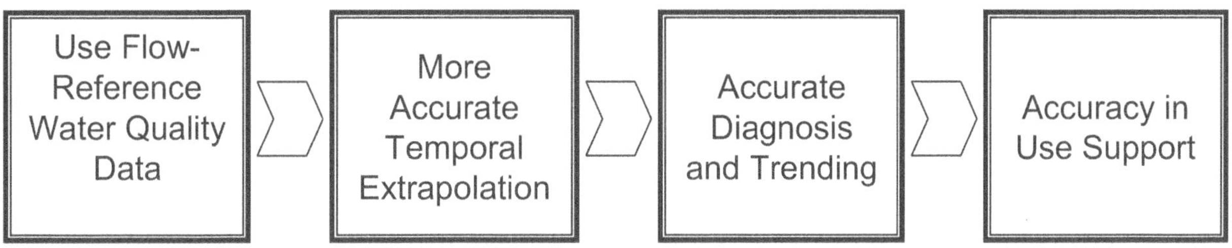

Figure 4-4. Benefits of Using Flow-Referenced Water Quality Data.

4.3.3 Standardize the Development of Assessment Units

The development of assessment units (AU) is critical to the outcome of waterbody assessments and to the conclusions drawn regarding the extent of water quality impairment. The delineation of waterbody AUs is based on a variety of factors for different states, including watershed boundaries, monitoring station locations, point-source discharge locations, tributaries, water diversion points, and land use. U.S. EPA guidance (2005a) envisions a system in which all state AUs would be uniquely identified and documented through a relational database, which tracks not only the location of the assessment unit, but also the designated uses for that unit, the waterbody assessment results, and the pollutants and sources of pollution for those units that are not attaining water quality standards. However, based on the information in the most recent Integrated Reports and assessment methodology documents, many states are still in the process of refining their AUs, and are not ready to provide the detail requested by U.S. EPA in an electronic, relational database format.

Currently there is no consistent spatial database of watershed delineations that is regularly used by all states. In order to achieve more consistency across the states in their definition of AUs, and to provide the relational database of information for each AU as requested by U.S. EPA, a national standardization of AU definition is needed. Based on a recommendation from the NRC Report (2001), states should develop a catalog of waterbodies based on the National Hydrography Dataset (NHD) for the purposes of defining state waters and designing sampling and assessment programs. The appendix to U.S. EPA's Integrated Report guidance (2005a) provides an introduction to a process by which states can define and link their stream segments (e.g., AUs) to the NHD. U.S. EPA (2005a) strongly encourages states to adopt the NHD reach addressing protocol for assigning stream segments, which will help to ensure that all waters of the state are being assessed. "The use of the NHD protocol for segment delineation provides powerful mapping and spatial analysis capabilities for all water quality characterization activities" (U.S. EPA, 2005a). Recent improvements to the NHD provide higher resolutions and watershed attributes than previous versions. More information on this tool can be found at http://nhd.usgs.gov/.

4.4 How States Determine Water Quality Standard Attainment

4.4.1 Develop Numeric Criteria to Translate Narrative Water Quality Standards

As discussed in Chapter 3.0, states have developed different assessment methodologies depending on the beneficial use and type of water quality indicator being evaluated. This approach is appropriate in that it ties the assessment methodology to the water quality standards established by the state. For those standards which are supported by numeric water quality criteria, there appear to be consistent approaches among many states (e.g., using the ten-percent rule) to measure compliance with these standards and support of designated use. These approaches, with sufficient documentation, can be easily evaluated and verified by interested parties.

However, for those standards which are not supported by numeric criteria, there are a wide variety of evaluative methods used to measure impairment. Most often, the indicators used to assess impairment with these narrative standards include biological, nutrient, and sediment data. Often the methods to determine impairment using these indicators require more subjective analysis, use of numeric thresholds that have not been publicly reviewed, and best professional judgment, thereby making the methods less transparent and auditable by stakeholders. It is recommended that states focus their efforts on developing numeric criteria for those water quality standards that are still narrative in nature or for which numeric thresholds are being used to determine impairment. Development of numeric criteria should be accomplished through a formal, public review and rulemaking process.

Though not within the scope of this project, the research team noted criticism (e.g., in Karr and Yoder, 2004; NRC, 2001; FWQC, 2002) of the states' current use of "general" designated uses, and the associated difficulty in developing ecologically meaningful water quality criteria for these vaguely defined uses (e.g., aquatic life use, cold-water fish assemblage). Karr and Yoder (2004) argue the following:

> A more reasoned approach employs tiered aquatic life uses in conjunction with a biological condition gradient to specify the ecological potential of each waterbody, thereby leading to more cost-effective decisions for pollution control and restoration. In

> setting designated uses for urban and agricultural landscapes, for example, expectations might be different from those in parks and reserves. Without tiered aquatic life uses, real differences among waters are not accounted for in the listing of impaired water bodies, an oversight that can lead to inappropriate TMDL decisions. (Karr and Yoder, 2004)

The development of numeric biocriteria will help the states better utilize and assess biological monitoring data and confidently determine aquatic life use support or impairment. "Using biological criteria and regional reference networks to improve chemical and physical criteria should resolve many inaccurate 303(d) listings" (Karr and Yoder, 2004). Assessments of waterbodies using biological community assemblage data "can overcome statistical difficulties posed by chemical sampling, avoid the cost of frequent chemical sampling, and measure actual resource condition rather than depending on chemical surrogates of biological response" (Karr and Yoder, 2004). The NRC Report (2001) recognizes the following:

> Biological assemblages respond to and integrate all relevant chemical, physical and biological factors in the environment whether of natural or anthropogenic origin. However, relying only on biological assessments does not allow precise enough determination of associated causes and sources of impairments to satisfy water quality management needs. (NRC, 2001)

The need for determining causes of impairment should therefore be considered when developing biocriteria.

Development of nutrient criteria is important to protect waterbodies sensitive to nutrient loadings and is particularly important for fulfilling requirements of CWA Section 314. The development of nutrient criteria should consider actual impairment of designated uses, which is best achieved on a site-specific basis.

To aid in the development of the assessment methodology used to measure water quality indicators against these criteria, the criteria should be designed to address not only magnitude, but also frequency and duration. The frequency component should be expressed in terms of a number of allowed excursions in a specific period. A criterion that includes magnitude, duration and frequency components will provide a clearly defined definition against which compliance will be measured, thus reducing uncertainty in the assessment outcome.

Based on the complex relationship between pollutant and biological community response, these numeric criteria should be developed based on regionally relevant reference conditions and designated use (FWQC, 2002). The NRC Report (2001) cautions that a waterbody lacking in biological integrity, as compared to a benchmark condition at minimally disturbed sites, should not necessarily be assumed to be impaired. "Measures of biological conditions (e.g., Index of Biotic Integrity [IBI]) inform society of the status of a water resource. But society must decide the desired designated use and then determine what level on the index numeric scale is, with reasonable certainty, likely to protect that designated use." The state should collect data to determine if the scale of a numeric standard should be based on ecoregion, watershed, or waterbody-specific reference conditions.

U.S. EPA should develop guidance to help states translate narrative standards to numeric criteria for waterbody assessment purposes (NRC, 2001). U.S. EPA (1990) provides national guidance regarding the development of biocriteria for surface waters. However, guidance that provides a standardized framework for developing biocriteria and capitalizes on more recent research and technology is needed (Yoder, 2004).

4.4.2 Allow for Weight-of-Evidence Approach to Determine Both Attainment and Non-Attainment of Beneficial Uses

While reviewing states' Integrated Reports and assessment methodology descriptions and interviewing state and U.S. EPA personnel, it became apparent that there is an inconsistency in the assessment approach described by several states and the assessment approach utilized by states to achieve approval of 303(d) list of impaired waters by U.S. EPA. Many states maintain that they use a weight-of-evidence approach to determine attainment of water quality standards, particularly those states that use biological data in conjunction with habitat and chemical data to determine support of aquatic life uses. Use of multiple indicators was highlighted in the NRC Report (2001), which recommends that states use biological indicators in conjunction with physical and chemical indicators to assess the condition of waterbodies, as they provide different and complementary types of information about the sources and extent of impairment. However, depending on the wording of the state's water quality standards, this kind of approach may not be accepted by U.S. EPA during its 303(d) list approval if different types of data indicate conflicting determinations of attainment versus nonattainment of water quality standards.

An Ohio study of over 2,500 sampling sites (cited in NRC, 2001) found, for example, that in 51.6% of samples, the results from biomonitoring and chemical monitoring agreed in determining attainment versus nonattainment of water quality standards. However, in 41.1% of the samples, the biomonitoring data indicated impairment, while the chemical data did not. In 6.7% of the samples, chemical data indicated impairment, while the biomonitoring data did not (Ohio EPA, 1990; Rankin and Yoder, 1990—cited in NRC, 2001).

U.S. EPA guidance (2005a) recommends that states apply the principle of independent applicability when making use support determinations with multiple data types. "U.S. EPA's policy on independent application is based on the premise that any valid, representative dataset indicating an actual or projected water quality impairment should not be ignored" (U.S. EPA, 2005a). This principle was defined for water quality assessment purposes in U.S. EPA's *Policy on the Use of Biological Assessments and Criteria in the Water Quality Program* (U.S. EPA, 1991). Application of this principle results in a precautionary approach to determinations of beneficial use attainment. Despite EPA guidance further clarifying the policy of independent applicability (Table 4-1), there appears to be some confusion and miscommunication regarding the application of this principle to "override" use support determinations obtained through a weight-of-evidence approach.

Table 4-1. Independent Applicability For Purposes of Water Quality Standard Attainment/Nonattainment Determinations.

Policy of independent applicability *says*:
- When evaluating multiple types of data (e.g., biological, chemical) and any one type of data indicates an element of a water quality standard is not attained, the segment should most likely be identified as impaired.
- If there is reason to doubt the nonattainment finding, re-evaluate all of the data sets to resolve discrepancies. In some cases this may lead to modification of applicable water quality standard to account for site-specific information.
Policy of independent applicability *does not say*:
- Always assume that a single sample result showing impairment outweighs all other data showing attainment.
- Accept all differences in data findings at face value.

From: U.S. EPA (2005a)

From review of U.S. EPA guidance (2005a), it appears that a weight-of-evidence approach (i.e., use of multiple indicators) to determine use support is allowed, and is even advocated in other federal guidance (U.S. EPA, 2005b). However, U.S. EPA requires that when conflicting results occur, the data be reviewed to determine if the differences in assessment results can be attributed to differences in quality of the datasets. For example,

> This may involve consideration of analytical methods, review of sampling techniques, and detailed assessments of datasets. When the differences are due to data quality issues, the independent applicability policy allows for resolving the differences by cleaning the data or weighing the higher quality dataset more favorably in the attainment decision. (U.S. EPA, 2005a)

In the case where biological impairment is due to physical (i.e., habitat) disruption and not chemical impairment, conflicting biological and chemical assessment results would be expected. This underscores the need to collect all three types of data (i.e., chemical, biological and physical), and to have clear decision rules when using multiple lines of evidence to determine attainment or non-attainment of water quality standards.

Under a true weight-of-evidence approach, "all available data would be evaluated using fundamental scientific principles concerning the assessment of data quality, sufficiency, and applicability, which would ensure that the best scientific analysis and the best available data are used to make impairment decisions" (FWQC, 2002). To provide greater consistency between the approach that many states want to use and the approach considered appropriate by U.S. EPA during its review of the state's 303(d) list, states should submit the decision rules of their methodology to U.S. EPA for review and comment. Decision rules should address the steps that the state will take to re-evaluate and verify data that produce conflicting use support determinations, and the weighting factors that will be used to determine the final assessment outcome. Allowing states to use a weight-of-evidence approach would provide an incentive for more data collection, of multiple data types, thus improving the state's understanding of the ecological condition of its waterbodies.

4.4.3 Use Statistically Based Evaluative Techniques for Comparing Chemical Data to Numeric Water Quality Criteria

Limited monitoring resources and reliance on outside data sources limit the states' ability to confidently determine compliance with numeric water quality criteria. Statistical inference procedures should be used on the sample data to test hypotheses about whether or not the condition of the waterbody meets the water quality criterion (NRC, 2001). This is particularly true for assessments based on non-toxic chemical data (i.e., conventional pollutants), for which most states allow a specified exceedance rate (usually 10%) to be considered non-impaired.

It should be noted that recent EPA guidance (2005a) recants the 10-percent rule recommendation for determination of non-attainment for conventional pollutants. Instead, EPA recommends that states use an allowable exceedance rate that is directly tied to the water quality criteria. For example, a 10-percent exceedance rate would not be appropriate for a water quality criterion expressed as an instantaneous or 1-day maximum not to be exceeded (e.g., toxic chemical pollutants). EPA argues that the 10-percent rule also is not appropriate for water quality criteria expressed as averages over a specified length of time. In other words, the analysis method chosen to evaluate attainment status support should be consistent with the water quality criteria against which the waterbody is being measured (U.S. EPA, 2005a).

Although EPA's new recommendation is consistent with others stated herein, many state water quality criteria have not been written with the assessment method in mind, and so may be difficult to translate into an assessment method that considers natural variability and uncertainty.

Although most states rely on the raw-score method to estimate the exceedance rate for chemical data, use of this method does not adequately address the issue of uncertainty and could misrepresent the number of impaired waters as determined by the state (NRC, 2001). Also, the raw-score method does not account for the number of samples, which can result in a large Type I error (the error of a false positive) (Smith et al., 2001). Some states use the raw-score method as a screening tool to develop a "preliminary" list of impaired waters, after which a more thorough assessment using statistical tests is used. This preliminary screening approach is supported by recommendations in the NRC Report (2001).

For water quality criteria that allow a certain frequency of exceedance, statistically based methods should be used to estimate this exceedance rate, as these types of methods take into account sample size and can provide determinations of impairment with a known confidence, while reducing Type I error. "Properly executed statistical procedures can identify the magnitude and direction of the possible [assessment] errors so that knowledge can be incorporated into the decisions made" (NRC, 2001). Examples of statistically based methods include the binomial method, upper or lower percentile estimations with associated confidence intervals (discussed in Gibbons, 2000), equivalence testing (discussed in McBride 1998, 1999) and the Bayesian Approach (discussed in Section 4.2.5).

An advantage of using statistical methods is to allow the states to analyze datasets with differing sample sizes and report the confidence and/or uncertainty associated with the result. U.S. EPA (2005a) discourages states from establishing rigid minimum sample sizes, and instead encourages the states to evaluate even the smallest datasets. U.S. EPA recommends that a state's assessment methodology provide decision rules for determining impairment when minimum sample sizes are not met but available data indicate a reasonable likelihood of exceedance of the water quality criteria. Because analysis of waterbodies with very small datasets can lead to inaccurate conclusions, using statistical methods and evaluating the uncertainty (i.e., Type I error, Type II error or Power) provides the states a way to qualify the confidence in results from assessments of waterbodies with little data.

When choosing statistical methods to measure compliance with water quality criteria, another consideration is the policy decisions implicit in choosing the null hypothesis, and controlling the Type I error rate (through selection of the significance level) and Type II error rate (through selection of minimum sample size). The nature of statistical hypothesis testing is to determine whether the evidence (data) is sufficient to refute the presumption (null hypothesis). Choosing a null hypothesis that a waterbody is "meeting water quality standards," along with a significance level of 0.05 (the Type I error of rejecting the null hypothesis when it is actually true is kept at or below 5 percent, a value often chosen in experimental studies), would result in finding that the waterbody is meeting water quality standards (i.e., not rejecting the null hypothesis) unless a great deal of evidence to the contrary is present. However, using a more precautionary null hypothesis (i.e., water quality standards are not being met) may result in a finding that the waterbody is impaired, *even with the same set of data*. For 303(d) listing purposes, the NRC Report (2001) recommends that the null hypothesis be that the waterbody is not impaired, whereas for delisting purposes, the null hypothesis should be that the waterbody is impaired.

U.S. EPA (2005a) recommends that when selecting statistical methods and decision rules for those methods, states should minimize the chances of making either of the following two errors:

- Concluding the segment is impaired, when in fact it is not (Type I error)
- Deciding not to declare a segment impaired, when it is in fact impaired (Type II error)

Minimization of these two errors requires careful balance and selection of the null hypothesis, significance level and minimum sample size. Smith et al. (2001) explains that the choice of an error rate should be a risk management decision based on explicit consideration of consequences (costs) of being wrong. Falsely concluding that a water segment is impaired results in unnecessary planning and pollution control implementation costs and may cause people to avoid use of that water, resulting in lost benefits to society. On the other hand, falsely concluding that a segment is not impaired may pose a risk to human health or to the services of the aquatic environment. Given limited resources available for programs of water quality improvement planning and implementation, it is important that waters that are truly impaired be identified (i.e., statistical power) (Smith et al., 2001). Calculation of statistical power (concluding a segment is impaired when it truly is impaired) would provide additional support for the particular statistical method chosen for assessment purposes.

4.4.4 Base Determination of Fully Supporting On Attainment of Minimum Water Quality

The format of the Integrated Report provides states a way to distinguish among waterbodies which they know are impaired, those which are not impaired, and those for which they do not have enough information to determine use support. Although most states have developed detailed assessment methodologies and decisions rules by which they can make a confident determination of impairment, only a handful of states have developed methodologies by which they can confidently make a determination of "fully supporting" of designated uses. It is recommended that states develop minimum water quality requirements that must be attained before a waterbody can be assessed as fully supporting a particular use, rather than just defaulting waters to "fully-supporting" based on a limited set of data. This will allow states to better distinguish high quality waters, which should be assigned to Integrated Report Category 1, versus waters which do not exceed water quality criteria and are not impaired. The NRC (2001) provides a recommended list of core water quality indicators which could be used for this purpose.

4.4.5 Develop Better Descriptions of the Linkage Between 305(b) and 303(d) Waterbody Assessments

Although many states have transitioned to the Integrated Report format, others are still producing separate 305(b) and 303(d) reports, often with a completely different evaluative process for the statewide assessment than for the waterbody specific assessments. Other states focus their Integrated Reports entirely on the waterbody specific assessments and do not provide adequate information on overall statewide or basinwide water quality. Florida's Integrated Report provides an example of a clear top-down assessment process in which the assessments performed for 305(b) purposes feed directly into the waterbody specific assessments performed for developing the 303(d) list. New York has a similar process, though it has not adopted the Integrated Report format. It is recommended that all states provide better descriptions of the linkage between the assessments for these two purposes, similar to that provided in the Florida and New York Integrated Reports. Providing this linkage will aid in the states' allocation of monitoring and assessment resources, and result in assessments that are performed in a more efficient and consistent manner. It also will ensure that assessments of general statewide and basinwide water quality will not recede in importance in the states' future Integrated Reports.

4.4.6 Provide More Detailed Descriptions of Assessment Methodologies to the Public

The states' descriptions of their assessment methodologies, as provided in the Integrated Reports, 305(b) or 303(d) reports, or separate assessment methodology documents, vary greatly in detail. Because these assessments determine which waterbodies have impaired designated uses and thus require a TMDL, it is imperative that the assessments performed by the states are transparent and auditable by stakeholders and the general public.

Generally, states have provided adequate descriptions of assessments that compare chemical data to numeric water quality. However, descriptions of assessments based on biological or physical data, in which no numeric criteria have been established, are often lacking in the detail necessary to critically evaluate the assessment or the data used. Without this type of rigorous documentation, a similar outcome from a repeat of the assessment may not occur, which would call the validity of the original assessment into question.

Several states (e.g., California, Montana) are providing additional documentation for each waterbody assessment beyond that of the Integrated Report and 303(d) list. States should provide thorough documentation for each waterbody assessment that will provide the public with information regarding the data used in the assessment, data verification procedures (QA/QC), data analysis procedures, and determination of designated use attainment. This information also will be useful for U.S. EPA in their review and approval of the state's 303(d) list. If states do not have the resources to develop documentation for each waterbody assessment, then effort should be made to produce a stand-alone, comprehensive assessment and listing methodology document, which would be available for public review and comment. This will ensure that specific assessment methodologies can be reviewed and verified by interested stakeholders.

4.4.7 Provide Descriptions of Specific De-Listing Assessment Methodologies

Although states generally provide descriptions of the requirements for de-listing a waterbody from the 303(d) list, very few states provide detail regarding the assessment methodology that would be used to evaluate whether or not the waterbody meets the de-listing requirements. Some states (e.g., California) have chosen to incorporate more rigorous proof that the waterbody is attaining its designated uses before being removed from the 303(d) list. U.S. EPA (2005a) recommends that states include in their Integrated Report those segment/pollutant combinations that have been removed from the 303(d) list since the last reporting cycle, along with the summary rationales for each delisting. It is up to the states, with EPA's approval, to determine the required burden of proof ("good cause") for de-listing. Adequate detail regarding the evaluative procedures that states will follow in reviewing use attainment for 303(d) listed waters should be provided to the public as part of the state's waterbody assessment methodology documentation.

4.5 Public Participation/Peer Review During Methodology Development

4.5.1 Provide a Forum for Public Input During Development of Waterbody Assessment Methodology

Although the states submit a draft 303(d) list of impaired waters to the public for review and comment, very few states allow the public to be involved in the development of the methodology that is used to create the 303(d) list. By bringing the public into the process at the end, many states invite criticism about their assessment methodologies, including the data used in the assessments. The states then often have to allocate resources to re-assess waterbodies using different data and evaluative techniques that are more appropriate to determine use support based on current conditions.

Public participation in technical discussions with the states during methodology development could result in early buy-in from the public and potential savings in both time and money after the publication of the draft 303(d) list. This public participation could be achieved through the drafting of an assessment and listing methodology document, similar to U.S. EPA's 2002 CALM guidance, that is published for review and comment. Another forum for public participation would be to allow the methodology development process to have oversight by a public technical advisory committee. Both of these types of public involvement are currently being used by some states to encourage public input early in the waterbody assessment process. Another option would be to adopt the assessment methodology as part of a rulemaking procedure into the state's water quality standards. This public process would provide maximum opportunity for stakeholder review and input, however, but would require a more formal U.S. EPA approval of the assessment methodology.

4.5.2 Submit Draft Assessment and Listing Methodology Document to U.S. EPA for Review and Comment

U.S. EPA Integrated Report guidance (2005a) notes that U.S. EPA will review the state's assessment methodology while determining whether to approve or disapprove the state's 303(d) list. U.S. EPA even encourages the states to make available their most recent waterbody assessment methodology prior to submission of their Integrated Report. Similar to the recommendation discussed above, achieving early buy-in from U.S. EPA regarding the waterbody assessment methodology could potentially save time and money after the publication of the draft 303(d) list. Without soliciting a formal approval, states should submit their draft assessment and listing methodology document to U.S. EPA for review and comment well in advance of developing the 303(d) list. It is envisioned that a well-developed assessment and listing methodology document that has had both public and U.S. EPA review would not change significantly between listing cycles, but would only be refined as data and evaluative techniques improve over time. Critical to this recommendation is for U.S. EPA to achieve greater consistency in its review and approval process between regions.

CHAPTER 5.0

SUMMARY AND CONCLUSIONS

5.1 What Do Current Waterbody Assessment Methods Look Like?

A critical first step to more effective integration of monitoring data and technically sound assessment methodologies into the state waterbody assessment and listing process is a critical evaluation of the current methodologies employed by the states. The research team evaluated waterbody assessment methodologies and documentation, and found both common and varying aspects elements among the state approaches. Conclusions about the major methodology components are summarized below.

What specific types of data are required to conduct waterbody assessments for listing purposes?

- Most states consider data from all readily available sources, though many (39) have specific data QA/QC requirements that must be met for data to be included in waterbody assessment for listing purposes.
- Thirty-two states have minimum sample size requirements in order to perform an assessment. Waterbodies which do not meet minimum data requirements are placed in Integrated Report Category 3 and targeted for future monitoring.

How is the assessment methodology tied to monitoring network design?

- Because of limited monitoring resources and the U.S. EPA requirement to consider all readily available data, the states have not designed monitoring networks specific to waterbody assessment for 303(d) listing purposes. Therefore, the states currently must allocate resources to determining the sufficiency and credibility of each data source during every assessment and listing cycle.

How does the assessment methodology extrapolate water quality over space/time where data are not available?

- Many states (12) have developed watershed-based assessment units (AUs), for which the outcome of an assessment at a single monitoring station can be applied to all waters within that assessment unit.
- For 305(b) purposes, several states (9) have developed a probability-based monitoring network designed specifically for statistical extrapolation of water quality conditions to all waters of the state. However, this extrapolation is not used for 303(d) listing purposes.

How does the assessment methodology determine water quality standard attainment or non-attainment?

- States have developed assessment methods which are specific to the designated use and water quality indicator being assessed.
- The majority of states (32) use the 10-percent rule to determine compliance with numeric water quality criteria for non-toxic pollutants, with 25 states using the raw- score method and 8 states using the binomial method to estimate the exceedance rate for a particular waterbody.
- For toxics, many states (18) use a one-in-three-year exceedance rule for impairment determinations, while others continue to use the 10-percent rule, utilizing both the raw-score and the binomial method.
- Biological assessments also are an important part of many states' (36) assessment of aquatic life use support; however, the decision rules (and documentation thereof) establishing reference threshold values for biological community scores vary greatly among the states.

Does the assessment methodology quantify uncertainty?

- Those states which use statistically based data evaluation techniques (13) quantify a confidence level associated with the outcome of the analysis. This addresses a portion of the uncertainty associated with the waterbody assessment.
- Most states adhere to the precautionary principle, a policy decision which favors Type I error (false positive or impairment) over a Type II error (false negative or attainment).

Was public participation/peer review sought when developing the assessment methodology?

- Most states solicit data from the public prior to performing assessments.
- All states publish their draft 303(d) list for public comment.
- Eighteen states published their draft assessment methodology for public comment prior to performing waterbody assessments.

5.2 What Does a Model Waterbody Assessment Methodology Look Like?

Using the strengths found in many of the states' assessment methodologies and information found in other guidance, the recommendations from this research have been summarized into a hypothetical "model" waterbody assessment methodology. Using the same methodology components which framed the description of current waterbody assessments, this model methodology would include the items listed below and discussed in the following paragraphs.

- Data Used in Waterbody Assessments
- Integrating Monitoring with Assessments
- Water Quality Extrapolation Techniques
- Determining Water Quality Standards Attainment
- Public Involvement

5.2.1 Data Used in Waterbody Assessments

The model assessment methodology would clearly outline the types of data that will be used by the state to determine attainment of water quality standards, including minimum data quantity and quality requirements. For data collected and submitted by outside (e.g., third-party) sources, the methodology would require that data QA/QC documentation be provided to ensure that the data adhere to the state's U.S. EPA-approved QAPP.

The methodology also would describe how data might be used to determine impairment if data quantity and quality requirements are not met (i.e., described by states as overwhelming evidence). This description would provide a clear connection between the assessment using the limited data and evidence of actual impairment of designated use.

Finally, the methodology would describe how both non-detects and outliers in the dataset will be handled during data evaluation. The use of special statistical techniques for handling non-detect data would be used where appropriate. A data validation review would be conducted on outliers, which would help determine the appropriate "weighting" of data collected under different conditions or under a more targeted monitoring objective than the ambient condition to be assessed.

5.2.2 Integrating Monitoring with Assessments

The model assessment methodology would build upon a state water quality monitoring network designed so evaluation of attainment of water quality standards is a guiding objective, both on statewide and site-specific scales. This monitoring network would be designed as part of a comprehensive, long-term monitoring and assessment program designed to most efficiently support all water quality programs. The long-term monitoring and assessment program would include an extensive biological monitoring component in order to support development of regional reference conditions and biocriteria, as well as routine bioassessments to determine support of aquatic life uses.

In order to reduce the uncertainty associated with using data from outside sources for waterbody assessment and listing purposes, the state would take steps to integrate state monitoring efforts with monitoring efforts of other entities that contribute data to the process. In addition, the methodology would require that documentation of monitoring objectives accompany all third-party data to ensure that the data reflect the aspect of condition being assessed by the state. Finally, the state's methodology would be updated if evaluative techniques emerged in the future that are more appropriate and scientifically defensible for assessing waterbodies using data from multiple sources.

5.2.3 Water Quality Extrapolation Techniques

For waterbody assessment and listing purposes, the state would organize available monitoring data into a geo-referenced database that is linked to the NHDPlus spatial database, thereby delineating watersheds of the state into a reasonable assessment scale. This watershed delineation would be used to segment waterbodies into AUs based not only upon homogenous hydrologic characteristics (i.e., stream order), but also on designated uses, land cover, land use, and locations of point and non-point pollution sources. For such defined AUs, determinations of compliance with water quality standards could be extrapolated throughout the unit, based upon the assumption of homogenous water quality.

5.2.4 Determining Water Quality Standards Attainment

The model methodology would provide specific decision rules and statistically based evaluative techniques to determine compliance with water quality standards which are specific to each designated use and water quality indicator being assessed. Ideally, these decision rules and evaluative techniques would be derived directly from the water quality standard itself, which would specifically describe the magnitude, frequency and duration for each water quality criteria that is to be met in order to be protective of designated uses. These numeric criteria would have been developed for each type of indicator used in the assessment, including chemical, biological and physical indicators.

For assessment purposes, the state would look at compliance with each water quality standard using one or more indicators, as data are available. For waterbodies assessed with multiple indicators, a true weight-of-evidence approach would be applied, in which conflicting results between indicators would be resolved by following a predetermined data validation process, and then by a determination of support or impairment of designated use based on weighted consideration of each indicator.

The state's Integrated Report would first provide a general assessment of all waterbodies of the state, and then provide a clear and logical connection between the statewide assessment and specific waterbody assessments. Each waterbody (or AU within each waterbody) would be assigned into one of the five Integrated Report categories, with documentation providing the rationale for each waterbody classification. Rationale would be provided for both addition and removal of waterbodies from Category 5 from the previous listing cycle. This rationale would include the data used in the assessment, the source and location of the data, associated data validation and QA/QC documentation, and the decision rule or evaluative technique used to determine support or impairment of designated use.

For waters to be listed in Integrated Report Category 1 (fully supporting), the methodology would establish a minimum set of core indicators that would have to indicate support of all designated uses. For waters to be listed in Integrated Report Category 5 (303[d] list of impaired waters), results from the waterbody assessment would need to indicate actual impairment of the designated use and evidence that the impairment was caused by a pollutant.

5.2.5 Public Involvement

The ideal waterbody assessment and listing methodology would be developed in an open, public process to allow for stakeholder input prior to the development of the list of impaired waters (303[d] list). Initial methodology development would include peer review by an expert panel of water quality managers, engineers and scientists. Then, the draft methodology would be submitted to the public for comment, as well as to U.S. EPA for review and comment prior to finalization. The methodology also would allow for periodic updates and public review of those updates as the methodology is refined in the future.

APPENDIX A

REPORT SEARCH INVENTORY

Appendix A. Report Search Inventory

State	AgencyName	AgencyWebsite	RelevantWebsites	ReportComments
US	EPA		http://www.epa.gov/owow/305b/	Link to National Assessment Database that has all submitted 2002 305(b) reports - contains summary information, not report itself (does not have info from 2004 reports) - may be useful for agency names and contacts
Alabama	Alabama Department of Environmental Management	www.adem.state.al.us	http://www.adem.state.al.us/WaterDivision/WQuality/305b/WQ305bReport.htm	Downloaded 2004 Integrated report and draft wq assessment methodology
Alaska	Alaska Department of Environmental Conservation Division of Water	http://www.dec.state.ak.us/water/index.htm	http://www.dec.state.ak.us/water/wqsar/waterbody/integratedreport.htm	Downloaded 2002-3 Integrated report and wq assessment methodology
Arizona	Arizona Department of Environmental Quality	http://www.azdeq.gov/index.html	http://www.azdeq.gov/environ/water/assessment/2004.html	Downloaded draft 2004 integrated report.
Arkansas	Arkansas Department of Environmental Quality	http://www.adeq.state.ar.us/default.htm	List of publications - Integrated report is under surface water section: http://www.adeq.state.ar.us/water/branch_planning/publications.htm	Downloaded 2002 integrated report and 2004 303(d) list
California	California Environmental Protection Agency	http://www.calepa.ca.gov/	Under surface water section: http://www.waterboards.ca.gov/quality.html This is 2002 305(b) report: http://www.waterboards.ca.gov/tmdl/305b.html Recent revision to 303(d) list: http://www.waterboards.ca.gov/tmdl/303d_update.html	Downloaded 2002 305(b) report - COULD NOT FIND 2004 305(b) report
Colorado				Downloaded 2004 305(b) report and 2002 listing methodology; no integrated report
Connecticut	Connecticut Department of Environmental Protection		2004 Integrated report: http://dep.state.ct.us/wtr/wq/calm/2004_calm.pdf	Downloaded 2004 integrated report.
Delaware	Delaware Department of Natural Resources and Environmental Control	http://www.dnrec.delaware.gov/	http://www.dnrec.state.de.us/water2000/Sections/Watershed/TMDL/305and303.htm	Downloaded 2004 integrated report.
Florida	Florida Department of Environmental Protection	http://www.dep.state.fl.us/mainpage/default.htm	http://www.dep.state.fl.us/water/monitoring/index.htm	Downloaded 2004 integrated report.
Georgia	Georgia Department of Natural Resources	http://www.gadnr.org/	http://www.gaepd.org/Documents/305b.html	Downloaded 2002 integrated report and January, 2004: Draft Georgia 2004 305(b)/303(d) List Documents. The Jan 2004 draft seems to be just a series of tables, not much text.
Hawaii	Hawaii State Department of Health		http://www.hawaii.gov/health/environmental/water/cleanwater/reports/cleanwater/reports/reports.html	Downloaded "FINAL 2004 LIST OF IMPAIRED WATERS IN HAWAII PREPARED UNDER CLEAN WATER ACT §303(d)" and other related info.
Idaho	Idaho Department of Environmental Quality	http://www.deq.idaho.gov/	http://www.deq.idaho.gov/water/data_reports/surface_water/monitoring/integrated_report.cfm	Downloaded 2002 draft integrated report, website says 2004 draft report will be available soon for public comment but specific release date not given
Illinois	Illinois EPA	http://www.epa.state.il.us/	http://www.epa.state.il.us/water/water-quality/	Downloaded 2004 305(b) report
Indiana	Indiana Department of Environmental Management	http://www.in.gov/idem/	http://www.in.gov/idem/water/planbr/wqs/quality.html	Downloaded 2004 Integrated report
Iowa	Iowa Department of Natural Resources	http://www.iowadnr.com/index.html	http://www.iowadnr.com/water/tmdlwqa/wqa/303d.html	The 305(b) and 303(d) reports consisted of some web text and mostly tables - not much narrative; found a section on the website for "2004 Impaired Waters List" and this had more relevant files - downloaded Iowa_2004FinalMethodology.pdf

Appendix A. Report Search Inventory

State	AgencyName	AgencyWebsite	RelevantWebsites	ReportComments
Kansas	Kansas Department of Health and Environment's	http://www.kdheks.gov	http://www.kdheks.gov/tmdl/basic.htm#Kansas%20303(d)	Downloaded 2004 303(d) list, listing methodology, and 305(b) report
Kentucky	Kentucky Division of Water	http://www.water.ky.gov/homepage_repository/About+the+Agency.htm	303(d): http://www.water.ky.gov/sw/tmdl/303d.htm 305(b): http://www.water.ky.gov/sw/swmonitor/305b/305b+Reports.htm	Downloaded 2004 303(d) and 305(b) reports.
Louisiana	Louisiana Department of Environmental Quality	http://www.deq.louisiana.gov/	http://www.deq.louisiana.gov/planning/305b/2004/	Downloaded Water Quality Inventory Integrated Report (Section 305(b) and 303(d) Reports) 2004 - NEED TO PRINT
Maine	Maine Department of Environmental Protection	http://www.maine.gov/dep/index.shtml	Integrated Report http://www.maine.gov/dep/blwq/docmonitoring/305b/index.htm#2004	Maine did integrated reports in 2004 AND 2002; Downloaded 2004 integrated report (6 sections)
Maryland	Maryland Department of the Environment	http://www.mde.state.md.us/	http://www.mde.state.md.us/Programs/WaterPrograms/TMDL/Maryland%20303%20dlist/final_2004_303dlist.asp	Downloaded 2004 integrated report.
Massachusetts	Massachusetts Department of Environmental Protection	http://www.mass.gov/dep/dephome.htm	http://www.mass.gov/dep/brp/wm/tmdls.htm	Downloaded 2004 integrated report.
Michigan	Michigan Department of Environmental Quality	http://www.michigan.gov/deq	http://www.michigan.gov/deq/1,1607,7-135-3313_3686_3728-12711--,00.html	Downloaded 2004 integrated report.
Minnesota	Minnesota Pollution Control Agency	http://www.pca.state.mn.us/	305(b): http://www.pca.state.mn.us/publications/reports/305b-2004integratedswreport.pdf Impairment assessment: http://www.pca.state.mn.us/water/tmdl/index.html#finaltmdl	Downloaded 2004 integrated report. Downloaded "Guidance Manual For Assessing the Quality of Minnesota Surface Waters for the Determination of Impairment 305(b) Report and 303(d) List"
Mississippi	Mississippi Department of Environmental Quality	http://deq.state.ms.us/MDEQ.nsf/page/Main_Home?OpenDocument	http://www.deq.state.ms.us/MDEQ.nsf/page/FS_SurfaceWaterQualityAssessments?OpenDocument	Downloaded 2004 305(b) report and 303(d) report.
Missouri	Missouri Department of Natural Resources	http://www.dnr.state.mo.us/index.html	http://www.dnr.state.mo.us/wpscd/wpcp/waterquality/	Downloaded 2004 305(b) report and 303(d) list and methodology.
Montana	Montana Department of Environmental Quality	http://www.deq.state.mt.us/index.asp	Integrated Report http://nris.state.mt.us/wis/environet/2004Home.html 2002 305(b) List and 305(b) waters database http://nris.state.mt.us/wis/environet/2002_305bhome.html Montana Environet: http://nris.state.mt.us/wis/environet/index.html	Downloaded 2004 integrated report.
Nebraska			http://www.deq.state.ne.us/Publica.nsf/Pages/WAT021	Downloaded "Methodology for Waterbody Assessment and Developing the 2002 Section 303(d) List of Impaired Waterbodies for Nebraska" & 2004 Integrated Report
Nevada			http://ndep.nv.gov/bwqp/303dlist.htm	Downloaded DRAFT Nevada's DRAFT 2004 303(d) Impaired Waters List and the 2002 303(d) Impaired Waters List.
New Hampshire			http://www.des.state.nh.us/wmb/2000-305b.pdf	Downloaded NH's 2000, 305(b) report
New Jersey	DEP	http://www.state.nj.us/dep/	http://www.state.nj.us/dep/wmm/sgwqt/wat/integratedlist/integratedlist2004.html	Downloaded 2004 integrated report. Downloaded Methods Document.
New Mexico	NMED/NMENV	http://www.nmenv.state.nm.us/index.html	http://www.nmenv.state.nm.us/wqcc/303d-305b/2004/index.html	Downloaded 2004 integrated report.
New York			http://www.dec.state.ny.us/website/dow/bwam/305b.html	Nick downloaded
North Carolina			http://h2o.enr.state.nc.us/tmdl/General_303d.htm#Downloads	Nick downloaded
North Dakota	North Dakota Department of Health, Division of Water Quality	http://www.health.state.nd.us/wq/	http://www.health.state.nd.us/wq/sw/Z2_TMDL/TMDL_Lists/B_TMDL_List.htm	Downloaded 2004 integrated report.
Ohio	Ohio EPA	http://www.epa.state.oh.us/	http://www.epa.state.oh.us/dsw/tmdl/2004IntReport/2004OhioIntegratedReport.html#download	Downloaded 2004 integrated report.

Appendix A. Report Search Inventory

State	AgencyName	AgencyWebsite	RelevantWebsites	ReportComments
Oklahoma	Oklahoma Department of Environmental Quality	http://www.deq.state.ok.us/	http://www.deq.state.ok.us/WQDnew/305b_303d/index.html	Downloaded 2004 integrated report.
Oregon	Oregon Department of Environmental Quality	http://www.deq.state.or.us/	Fact sheets and reports: http://www.deq.state.or.us/pubs/factsheets.asp Integrated report as database: "Final 2002 303(d) Database Search Choices Page" http://www.deq.state.or.us/wq/wqldata/searchchoice02.htm	Downloaded 2004 305(b) report and listing methodology for 303(d) and 305(b); search for integrated report returns "Final 2002 303(d) Database Search Choices Page" - can use map to see waterbodies and listing status for various constituents in selected wat
Pennsylvania	DEP	http://www.dep.state.pa.us/	http://www.dep.state.pa.us/dep/deputate/watermgt/Wqp/WQStandards/303d-Report.htm	Downloaded 2004 integrated report and assessment methodology. List of categorized waters were in multiple files with some html - did not download or print these.
Rhode Island			http://www.dem.ri.gov/programs/benviron/water/quality/surfwq/index.htm	Downloaded "STATE OF RHODE ISLAND 2004 303(d) LIST LIST OF IMPAIRED WATERS FINAL May 2005"; Downloaded 305(b) report
South Carolina	South Carolina Department of Health and Environmental Control	http://www.scdhec.gov/	http://www.scdhec.gov/water/html/tmdl.html#303d	Downloaded 2004 integrated report.
South Dakota	South Dakota Department of Environment and Natural Resources	http://www.state.sd.us/denr/denr.html	http://www.state.sd.us/denr/document.htm#Water%20Quality%20&%20Watershed%20Protection:	Downloaded 2004 integrated report.
Tennessee	Tennessee Department of Environment and Conservation	http://www.tennessee.gov/environment/	http://www.state.tn.us/environment/wpc/publications/#303d04	Downloaded 2004 305(b) report and 303(d) list.
Texas		http://www.tceq.state.tx.us/	2004 Methodology for Developing the Texas List of Water Bodies Not Meeting Surface Water Quality Standards	Downloaded 2004 Methodology
Utah	DEQ	http://www.deq.utah.gov/	http://www.waterquality.utah.gov/TMDL/index.htm 305b: http://www.waterquality.utah.gov/documents/DOC_RULE.HTM	Downloaded 303(d) and 305(b) reports.
Vermont			http://www.anr.state.vt.us/dec/waterq/wqdhome.htm	Downloaded "2006 VERMONT SURFACE WATER ASSESSMENT METHODOLOGY"
Virginia	Virginia Department of Environmental Quality	http://www.deq.virginia.gov/	http://www.deq.virginia.gov/wqa/ir2004.html	Downloaded 2004 integrated report.
Washington	Washington State Department of Ecology	http://www.ecy.wa.gov/ecyhome.html	Water quality assessments home page: http://www.ecy.wa.gov/programs/wq/links/impaired_wtrs.html#305b 2004 Integrated report consists of webpages (not in PDF) and includes database and query tool: http://www.ecy.wa.gov/programs/wq/303d/2002/2002-index.htm	Website says Washington's Water Quality Assessment for 2004 has been submitted to EPA as an "integrated report"; Document provided on website through webpages, not in one file - did not see listing methodology in the integrated report webpages; Downloaded 2002 report
West Virginia	West Virginia Department of Environmental Protection	http://www.dep.state.wv.us/	http://www.dep.state.wv.us/item.cfm?ssid=11&ss1id=720	Downloaded 2004 integrated report.
Wisconsin	DNR	http://dnr.wi.gov/	http://dnr.wi.gov/org/water/wm/watersummary/305b_2004/index.htm	Downloaded 2004 integrated report.
Wyoming			http://deq.state.wy.us/wqd/watershed/index.asp#Assess	Downloaded 2004 305(b) & 303(d) State water quality assessment report and waters requiring TMDL's. Downloaded, "Wyoming's Method for Determining Water Quality Condition of Surface Waters"

APPENDIX B

TEMPLATE EVALUATION SURVEY

Appendix B
Template Evaluation Survey

Date (of evaluation):____

State: ____

Report Title:____

Type of report:__________________________ Month/Year:____________

Supplemental info/reports referenced:____________________________

____

Note: Provide report page number where information was found for each response below.

General

1. Is method transparent and easily understood via description in Integrated Reports (i.e. is method description adequate to allow calculations to be reproduced if desired)?
   (Page ______)

2. Is it clear under which circumstances the method will be used?
   (Page ______)

3. Does methodology incorporate monitoring network design considerations (i.e. is data analysis tied to monitoring framework)?
   (Page ______)

4. Does the method require statement of DQOs and/or information goals?
   (Page ______)

5. How does method extrapolate water quality over time/space where data are not available?
   (Page ______)

6. Is there sample-size requirement when drawing conclusion from the analysis results?
   (Page ______)

7. What data management protocols are needed to support method?
   (Page ______)

8. What documentation is available to describe method and its application?
   (Page ______)

Appendix B
Template Evaluation Survey

9. Is method easy to conduct, understand and explain to others?
   (Page ______)

10. Is method modified in any way depending on waterbody type or specific beneficial use?
    (Page ______)

11. How does method specifically determine Water Quality Standard attainment status?
    (Page ______)

12. Does method employ modeling?
    (Page ______)

13. Was any public participation or input sought when developing method?
    (Page ______)

14. Has method undergone peer review and is it generally accepted by current practitioners?
    (Page ______)

15. Does the method reference EPA statistical guidance documents?
    (Page ______)

16. Does method match "duration" and "frequency" components of the water quality standard?
    (Page ______)

17. What method/process does state employ when minimum data quantity/quality conditions are not met?
    (Page ______)

18. Is method answering the question of "how often" (i.e. frequency) or "how much" (i.e. amount) and is this appropriate for data analysis objective?
    (Page ______)

19. Which states use this method?
    (Page ______)

20. How long has method been used (i.e. what is track record)?
    (Page _____)

21. What software is required to conduct calculations and how much does it cost?
    (Page _____)

22. What are notable strengths and weaknesses of method?
    (Page _____)

Data

23. What kind of data is used/required for method? (i.e. probabilistic sampling, narrative, random grab samples)
    (Page _____)

24. Does method integrate biological monitoring data?
    (Page _____)

25. If method uses bioassessment data to make assessment determinations – are threshold values for water impairment determination addressed in methodology?
    (Page _____)

26. Does method integrate geo-referenced data?
    (Page _____)

27. Does method use only state monitoring data or other data (i.e. permittee data, volunteer monitoring data)
    (Page _____)

28. If non-state monitoring data are used, how does the state solicit data from other entities?
    (Page _____)

29. Does method consider non-monitored data? (i.e. data from mass-balance calculations or predictive modeling)
    (Page _____)

Appendix B
Template Evaluation Survey

30. Does method consider non-numerical or ancillary data? (i.e. climatic, ecological, land use information)
(Page _____)

31. Does method consider QA/QC of data?
(Page _____)

32. How is meta-data considered in the assessment methodology?
(Page _____)

33. How does method handle non-detects in data record?
(Page _____)

34. Does method require graphical displays of data?
(Page _____)

35. Does the method give more weight to higher quality data?
(Page _____)

36. Does the method restrict the age of data that may be used?
(Page _____)

Statistical Methods

37. How does method quantify uncertainty?
(Page _____)

38. How does the method deal with "natural background"?
(Page _____)

39. Does method use statistical hypothesis testing?
(Page _____)

40. If hypothesis testing, what are parameters and null hypothesis?
(Page _____)

Appendix B
Template Evaluation Survey

41. If hypothesis testing used – what is assumed distribution?
    (Page ______)

42. If hypothesis testing used – what significance level is used to reject null hypothesis?
    (Page ______)

43. Is method chosen to minimize both Type I and II error?
    (Page ______)

44. Are distribution-free, non-parametric statistics used? Which ones?
    (Page ______)

45. Are estimation methods used in lieu of, or in conjunction with, hypothesis testing?
    (Page ______)

46. Does the method consider data distributions other than normal or lognormal, such as gamma?
    (Page ______)

Level of Effort

47. How many personnel hours were spent in developing and performing the assessment methodology?
    (Page ______)

48. How many personnel were involved in developing and performing the assessment methodology? For how long?
    (Page ______)

APPENDIX C

STATE EVALUATION MATRIX

Appendix C. State Evaluation Matrix

State	How does method specifically determine Water Quality Standard attainment?	What specific type of data are required to conduct waterbody assessment method? Is there a minimum sample size?	Are there specific QA/QC requirements for data used in the assessment? Is there a credible data law in effect?	If data do not meet specific QA/QC requirements, how is the data used in the assessment?	How is the assessment method tied to monitoring network design (i.e. is there a specific monitoring program for 305(b) or 303(d))?	Does assessment method quantify uncertainty?	How does method extrapolate water quality over space/time where data are not available?	Was public participation/peer review sought when developing method? Is third-party data solicited from public?	Notable strengths of method
Alabama	Assessments differs per use. For conventionals, generally uses 10% exceedance rule - utilizing the binomial procedure at 90% confidence. Also relies heavily on biological assessments, as well as fish consumption advisories. Described assessment methodology appears to be the same for each different use (i.e. looking at fish consumption advisory for DW or Rec use).	Use readily available data, generally gathered in the previous six years. Includes water chemistry, land use, physical, habitat evaluations, biological and bacteria. Very specific minimum data requirements and data collection requirements depending on both use and waterbody type (i.e. 10 samples for water chemistry). Each waterbody must have several types of data (i.e. chemical, biological, fish tissue, bacteria, etc). Wadeable Multi-Habitat Bioassessment (WMB) data are weighted more heavily than chemistry, though must be in combination with habitat assessment.	Data must be collected through specific approved SOPs to meet minimum data requirements. This includes approved QAPP and study plan. No credible data law.	Data not meeting minimum data requirements are not used. Waterbodies which do not have minimum data requirements are not assessed further and placed in Category 2.	Rotating basin monitoring done for several different monitoring purposes.	Yes, 90% confidence level associated with binomial method and allowable exceedances for chemical constituents.	Geo-referenced assessment units (same designated use) and monitoring units (same land cover/use) based on watershed and sampling stations.	Public notice for data solicitation, draft methodology, and draft 303(d) list.	Assessment unit/monitoring unit designations for waterbody segmentation/assessment. Multiple types of data required before impairment determination.
Alaska	Assessment methodology based on documentation of "persistent" exceedances of water quality criteria, and BPJ.	Assessments based on "credible" data. water chemistry data primarily used, but also consider photos and other documentation of water quality impairment. No minimum data requirements, but won't list on one exceedance.	Credible data should have sampling plan and QA/QC procedures consistent with state standards.	Data not used, though specific requirements are not listed, and are subject to BPJ.	No monitoring network specific to 305(b) and 303(d)	No	No data extrapolation	Integrated report and 303(d) list are submitted to public for comment	
Arizona	Uses weight-of-evidence approach to assess waters. At least three samples of "core parameters" must be sampled and not exceed criteria to designate water as "attaining" designated use. Assessment methodology written into standards, and is different per water quality criteria, for example: toxics > 2 exceedances in 3 years; human health: binomial method (90% confidence level).	All readily available, credible and scientifically defensible monitoring data, includes physical, chemical, biological and toxicological data. Use third-party data. Most assessment based on water chemistry data. Data must be spatially (>200 m) and temporally (>7 days) independent. Take into account field precision. Minimum sample size of 20 collected during 3 or more events.	Both QAP and SAP requirements for data to be deemed credible, as well as approved laboratory. Arizona has a credible data law.	Data not used.	fixed station monitoring for baselines and trends, as well as targeted rotating basin monitoring, however this monitoring serves purposes other than just Integrated Report.	Yes, through binomial method for certain water quality criteria.	stream "reaches" are defined which contain one or more sampling station - assessment applies to whole reach - georeferenced.	Integrated report and 303(d) list are submitted to public for comment	Minimum "core" parameters to determine if waters are "attaining" = precautionary principle.
Arkansas	Assessment differs slightly per water quality parameter and ecoregion, but generally uses 10 to 25% exceedance rule (raw score method). Violation of narrative standards based on BPJ.	Use both state and third-party data collected within the last 5 years. Minimum of 12 samples for stream data, 4 samples for lakes and bacteria.	Require that third-party data be collected under a QAPP equal to or more stringent to state's, but verification of this is not evident.	Not used. Evaluated data can be used to determine "support".	State has ambient, roving and special studies monitoring, but not specifically tied to Integrated Report.	No	No data extrapolation	Not mentioned	
California	Assessment method differs per criteria type. Generally binomial method applied. If number of exceedances greater than allowed exceedances per binomial tables, then impaired. Fact sheets generated for each waterbody/pollutant impairment analysis.	All readily available data are considered, including third-party data. Data actually used to assess water quality standards attainment should be data that quantified and qualified. Information that is descriptive, estimated, modeled or projected may be used as ancillary lines of evidence for listing or delisting decisions. Minimum sample size requirement per criteria, different sample size for listing versus delisting.	Yes, QA/QC requirements for data used to determine impairment. No credible data law in CA	Data that does not meet QA/QC requirements can not be used for listing alone.	No specific monitoring network. Data is solicited.	Uncertainly is built into the binomial method - the minimum number of exceedances is chosen so that Type I and II error is <= 20% for listing toxics and conventional pollutants. When delisting toxic pollutants, Type I and II error rates are <=10%. When delisting conventional pollutants, Type I and II error rates are <=20%.	Data are not extrapolated	Yes, public involved in review and comment of draft policy (methodology).	Fact sheets provide basis for all listing decisions. Recognizes and tries to minimize uncertainty by using binomial method.
Colorado	Attainment based on numeric wq criteria. For chronic: if 85th% of data exceeds criteria, then impaired. For acute, if more than 1 exceedance in 3 years, then impaired. Bioassessments are considered on a case-by-case basis, evaluated against expected (i.e. reference) condition. Impairment due to sediment uses specific assessment outlined in state guidance.	Use water column (grab), toxicity tests, physical and biological assessments and modeling data, but only chemical data could be used for 303(d) methodology described. State collects data on a rotating basin system and solicits third-party data.	Data should be supported by QA/QC documentation, but not required to be submitted. No credible data law.	Waterbody may be placed on State's M&E list (Category 3) instead of 303(d).	Rotating basin monitoring done for several different monitoring purposes.	No, though comparison of 85% takes natural variability into account.	No data extrapolation	Yes, draft methodology, data solicitation, and draft 303(d) all go to public for comment.	
Connecticut	Assessment applies weight-of-evidence approach when using multiple data types (only one data type needs to indicate impairment). Use biological community assessments (compared to reference) for aquatic life use (standards are narrative, not numeric) in streams with specific decision criteria applied to classify waterbody as supporting, threatened, not supporting. Use BPJ for lake and estuary assessments. Decision criteria also used for fish consumption use, shellfishing, recreation, etc...using multiple types of data. For conventaionals, use 25% exceedance rule, toxics use 10% exceedance (raw score method).	All readily available data are considered, including third-party data. Includes chemical, physical and biological data types, both numberic, modeled and qualitative.	"Monitored" assessments based on "sufficient and credible" data, i.e. quantity and quality of information can result in reproducible results, with data < 5 years old. State or USGS data considered "monitored", as well as third-party data supported by a QAPP.	Considered "evaluated" assessment. Not used for 303(d) list.	Probabilistic monitoring underway in order to achieve more monitored waters - not specific to 305(b) or 303(d) as third party data is used.	No	Probabilistic monitoring program	Yes, through solicitation of data and draft 303(d) list	
Delaware	Assessment differs per use and wq parameter, as well as type of data. Example: continuous sampling average daily DO uses 10% exceedance, daily minimum uses 1% exceedance. Grab DO data assessed using nonparametric estimate of 10th% (10 or more samples, 90% confidence). If few data available, 2 or more exceedances = impairment. Nutrients (10 or more samples): nonparametric estimate of 90th% compared to standard, also looks at algal blooms, trends. Null = attainment. Aquatic life use: biological and habitat assessments compared to reference. Example: biological @ 67% of reference = Cat 5. ERES waters: use nonparametric trend tests (account for seasonality) at 95% confidence. Ammonia & temp: 2 or more exceedances.	All readily available data are considered, must be in electronic format. Data used from last 5 years used in assessments, last 10 years for trends. 10 sample size minimum for nonparametric estimates of percentiles.	State data used if collected and analyzed according to DNREC QAPP. Third-party data used if QA/QC is acceptable. State rarely gets third-party data.	Waterbody placed in Cat 3.	State monitoring of >200 sites used in 305(b) assessments. Extensive biological/habitat assessments at 332 sites in early 1990's.	Yes, via confidence levels on nonparametric estimates of percentiles and trends.	No data extrapolation	public comment on Integrated Report and 303(d) list.	Nonparametric estimates of percentiles and trend testing (to assess degradation of outstanding waters).

State	How does method specifically determine Water Quality Standard attainment?	What specific type of data are required to conduct waterbody assessment method? Is there a minimum sample size?	Are there specific QA/QC requirements for data used in the assessment? Is there a credible data law in effect?	If data do not meet specific QA/QC requirements, how is the data used in the assessment?	How is the assessment method tied to monitoring network design (i.e. is there a specific monitoring program for 305(b) or 303(d))?	Does assessment method quantify uncertainty?	How does method extrapolate water quality over space/time where data are not available?	Was public participation/peer review sought when developing method? Is third-party data solicited from public?	Notable strengths of method
Florida	305(b) - data from Tier I probabilistic monitoring program compared to "target" to determine statewide use support. 303(d) - For more specific basin assessments, uses decision processes for listing impaired waters outlined in "Impaired Surface Waters Rule" (IWR). Generally binomial statistics are used to compare water quality to standards with no more than 10% exceedance. Uses 80% confidence level for "planning list" and 90% confidence for "verified list". Nutrients: annual threshold or "increase of over 50% during 2 consecutive years". Biological Indices used and compared to thresholds.	All readily available is used, specifically quantitative biological data, water chemistry, nutrient impairment data, beach advisory, shellfish harvesting classifications and fish consumption advisories. State has extensive monitoring data used (Tier II). Data must be <10 years old for planning list, <7.5 years for verified list. Planning List minimum sample size requirements typically 10, unless repeated excedances (e.g. 3 wq or 1 toxic exceedance in 3 years for aquatic life use). Verified List requires 20 samples. State currently working with EPA to allow option to use older data (if representative) and smaller sample sizes (if credible overwhelming evidence indicates impairment).	All data used, including those provided by a third-party, must meet the requirements of the Department's QA rule and make available their supporting QA/QC metadata. Chapter 403.067, Florida Statutes requires that Florida's assessment methodology be based on scientifically valid methods and credible data.	Any data not meeting the necessary QA/QC requirements are flagged and not used in the final assessment process for the 303(d) list. These data are stored separately and their status, along with the reason(s) for being discarded, is made known to the public. Data without adequate or demonstrated QA may be used to place waterbodies on the Department's "Planning List" of potentially impaired waters and prioritized for further monitoring/study.	Tier I network (probability-based sampling) to determine statewide water quality status. Tier II monitoring data used along with others for actual basin assessments and 303(d).	Yes, through binomial method and confidence levels (80% for planning, 90% for verified).	No extrapolation.	The Impaired Waters Rule was developed over a 2-year period, with the assistance of a highly qualified technical advisory committee. This rule (i.e., the surface water assessment methodology) was written in response to a requirement in the 1999 Florida Watershed Restoration Act (403.067, Florida Statutes). The committtee meetings and all subsequent rule development meetings were publicly-noticed. Third-party data are actively solicited. The Department holds a series of public meetings to describe the purpose of the TMDL program, the watershed assessment process, and the role of the public and other stakeholders in the area proximal to the waters being evaluated. Public can submit data, and provide comment on Integrated Report and 303(d) list.	Planning and Verified 303(d) list screens data first and then uses more rigorous analysis to determine impairment. Assessmenty method is a transparent process that makes clear the data and information needed to make an impairment call.
Georgia	Assessment differs per water quality parameter. Fecals: 2 exceedances = not supporting. Metals: >25% = not supporting. WET testing: 1 exceedance = not supporting if instream data corroborate. Conventionals: >25% = not supporting. Fish consumption: no consumption = not supporting, also mercury thresholds.	Specifically used data from state monitoring programs, state federal and local governments, Clean Lakes projects, and three electric utility companies. No minimum sample size.	No specific QA/QC or credible data law.	NA	State monitoring program for several purposes, not just 305(b) or 303(d).	No	No data extrapolation	No.	
Hawaii	Tiered assessment process, depending on water quality parameter and sample size. For conventionals, uses comparison of geo mean to geo mean standard and lists on one exceedance, providing minimum sample sizes and sample distribution are met. Also uses a 10% and 2% exceedance rule for some parameters based on wq criteria. Habitat and biological assessments used for narrative criteria.	Use readily available data, though assessment methods refer to numeric, monitored data. Minimum sample sizes are required for listing, 10 for geo mean comparisons, 100 for 10% criteria and 500 for 2% criteria.	Yes, QA/QC documentation must be available upon request, as well a photos and descriptions of sampling locations (including assessment of surrounding land use and habitat). No credible data law.	Not used for listing, but considered higher priority for monitoring.	State has routine monitoring, but not specific for 305(b), 303(d).	No	No data extrapolation	Data is solicited from public for 303(d).	
Idaho	Differerent assessment methodology per use. Some waters do not have designated use, but are assessed based on presumed aquatic life and possibly recreation. IDEQ states that to assess as not supporting of use, must demonstrate that: (1) data show measurable and adverse change to beneficial use, (2) adverse change must be linked to a pollutant, and (3) pollutant must be linked to human-caused action and vice versa to demonstrate full use support. Bioassesment and BPJ used to assess compliance with narrative criteria. For numeric criteria, >10% exceedance = impairment. For conventionals, if <10% exceedance but measurable adverse effect = impairment, otherwise not impaired. Special considerations for temp and spawning periods. Main assessment tool is multimetric index assessment that integrates bio, phys and chem data.	WBAG is designed to assess state's BURP data, but will consider readily available data from other sources. Data classified as Tier I, II or III based on scientific rigor and relevance. Only Tier I (BURP-compatible) used for 303(d) listings. Tier I data considered BURP-compatible based on sampling protocol, electronic format, and integration into multiletric indices. BURP data include macroinvertebrates, fish, algae and habitat. Minimum of 2 data types needed for listing. Outside data must have been analyzed and conclusions drawn, or DEQ will not use for listing.	Must have QA/QC to be considered Tier I data.	Not used as Tier I, ths not used in 303(d) listings. Can be used for other assessment purposes (i.e. 305(b)).	Assessments rely heavily on RBP data collected through state's Beneficial Use Reconaissance Program (BURP).	No	Uses statewide geo-referenced system (WBID) as the foundation for extrapolating results, based on USGS 8 digit HUC. WBID stratification system uses stream order and land use to classify waterbodies for extrapolation. If 2 or more monitoring sites in waterbody, defers to "lower" support status determination.	WBAG published for public comment. Data solicited from public, 303(d) draft list is made public.	WBID waterbody identification program. Extensive biomonitoring program (BURP). Good documentation of bioassessment process (mutimetruc indices).
Illinois	Assessment differs per use. Aquatic life: based on biological assessments (indices) compared to thresholds, water quality standards are secondary in this assessment. For water chemistry and bacteria, uses 10%-25% exceedance rule (raw score) (TSS must be <50% for bacteria evaluation). Also has methodology for determining potential causes of impairment, as well as potential sources. For drinking water use: both 10% rule and one exceedance of MCL = impairment.	Use all readily available data. "Monitored" data considered to be site-specific chemical, physical and biological monitoring data no more than 5 years old - this is used for listing. "Evaluated" data is land use, older monitoring data, and volunteer data. No minimum sample sizes.	Yes, state has its own Quality Management Program. Third-party data is required to submit a QAPP. No credible data law.	Data is not used. State even highlights specific datasets that were not used in assessments because of data quality problems.	State has several water quality monitoring programs - data from all are used in 305(b) and 303(d), though not specifically designed just for these purposes.	No	Wadeable streams: 10 miles up/down, non-wadable streams: 25 miles up/down, large rivers: 50 miles up/down.	Data solicited from public. Appears public not involved until TMDL development.	Specific methodology for identification of potential causes and sources.
Indiana	Assessment differs per use and wq parameter. Conventionals: 10% exceedance rule. Toxics: case-by-case (BPJ). Nutrients: stream response (BPJ). Biological communities: compared to threshold values. Also look at "adverse effects". Sediment toxics: compared to 75%.	Only uses state data from their rotating-basin and fixed station monitoring program - chemical, biological, habitat and bacteria data are collected. Uses IBI's from state's probability-based survey (fish community) of surface waters. Uses lake trophic state index. No minimum sample size.	Only use state data, which is collected in accordance with EPA approved QAPP.	NA	Data from state's probability-based network and targeted monitoring network provide data for assessments, but monitoring done for more purposes than just Integrated Report.	No	305(b) - Probability-based assessments are extrapolated for entire watershed. Site-specific data needed for 303(d).	Draft 303(d) list is published for public comment.	

Appendix C. State Evaluation Matrix

State	How does method specifically determine Water Quality Standard attainment?	What specific type of data are required to conduct waterbody assessment method? Is there a minimum sample size?	Are there specific QA/QC requirements for data used in the assessment? Is there a credible data law in effect?	If data do not meet specific QA/QC requirements, how is the data used in the assessment?	How is the assessment method tied to monitoring network design (i.e. is there a specific monitoring program for 305(b) or 303(d))?	Does assessment method quantify uncertainty?	How does method extrapolate water quality over space/time where data are not available?	Was public participation/peer review sought when developing method? Is third-party data solicited from public?	Notable strengths of method
Iowa	For physical, chemical and bacteria data, use 10% exceedance rule (raw score). For toxics, >1 violation in 3 years. For aquatic life, uses biological index metrics compared to thresholds to determine impairment. Uses trophic state index for lake assessments. Bacteria in streams: geo mean of at least 10 samples > geo mean standard, or >10% exceed max standard. Bacteria in lakes: geo mean of 5 samples > geo mean standard, or beach closure. Fish consumption advisory = impairment. Occurrence of 1 pollutant caused fish kill in recent 3 years = impairment. Also compared mussels index from 1980s to current, if declined by >50% then impaired.	Gather all readily available data: physical, chemical, biological, including state, third-party data and modeled data, fish kill reports, BPJ from biologists (aquatic life use). However, only "monitored", site-specific data less than five years used for listing. Data must meet "data completeness" (i.e. minimum sample size) guidelines and "credible data" guidelines to be used in assessment. Generally only use data from most recent three years. Exception for "overwhelming evidence".	"Credible data" defined as data collected under a scientifically accepted sampling and analysis plan, including QA/QC procedures. Iowa has a credible data law.	BPJ data, data older than 5 years, data that do not meet "completeness" guide, data not meeting QA/QC, and habitat assessments are not used for 303(d) listing, though can be used in 305(b) and placed in other categories.	Data used from various state monitoring programs, not specifically designed for 305(b), 303(d).	Discussed in relation to sample size requirements, but not formally quantified.	No data extrapolation for 303(d)	Methodology submitted for public comment.	Data completeness guidelines and flowchart showing how water quality data are used to assess waterbodies.
Kansas	Assessment differs slightly per designated use/water quality parameter, but generally uses 10% exceedance rule, using the binomial method (screen first with raw score). Nitrate in DW has 0% exceedance rule, with other DW parameters allowed 50% exceedance. Use thresholds for nutrients in lakes and wetlands to list. Uses biological metric to list (compare to reference). Also designated waters which "almost" exceed 10% in the last year of sample as impaired (degrading trend). Uses a parametric statistical method in order to "prioritize" TMDLs.	Chemical and biological data from state monitoring, as well as other "readily avaible data" from third-parties. Accept modeling data, fish consumption advisories, etc. Generally minimum sample size of 12 for chemistry, 3 for biology.	State data undergoes rigorous QA/QC, and makes up majority of data used for 303(d) listings. Specific QA/QC not required for third-party data, though data must be "verifiable". QA/QC and metadata should be submitted with third-party data.	Data not verifiable won't be used for 303(d), but may be used in planning future monitoring.	State has extensive monitoring network for streams and lake chemistry, fish tissue and biological data collection. "Assessment units" have been developed as the contributing watershed for each station.	Yes, 90% confidence level chosen for binomial procedure and establishment of minimum sample size.	Samples from single station in each "assessment unit" apply to entire unit, as water from each assessment unit drain to that station. No temporal data extrapolation is done.	Yes, both methodology and 303(d) list submitted to public for comment.	Statistical procedure to determine impairment, as well as prioritize TMDL development. Initial "screening" procedure.
Kentucky	Assessment depends on use. Aquatic life: conventionals uses the 10% exceedance rule (raw score method) and an exceedance of 11% to 25% results in partial support for those ambient water quality stations with a minimum of 12 monthly data sets; metals uses 10% rule for chronic, and 1/3 years for acute; biological data uses metrics compared to thresholds. Recreation bacteria uses 20% exceedance. Fish consumption uses advisories and threshold for mercury. Lake assessments use both quatitative and qualitative data outlined in a decision matrix.	Use all readily available data that has reliable QA/QC. Chemical, physical, and biological data from site-specific surveys less than 5 years old considered "monitored". Older data and anecdotal information considered "evaluated". Citizen data not used directly for 303(d) listing, however, beginning with the 2008 305(b) cycle the commonwealth will consider volunteer monitored data that has proper QA/QC through a cooperative process between DOW and volunteer groups. Minimum sample size is required for pathogen monitoring	Require reliable QA/QC, but no specific requirements.	All data not having approved SOPs and QAPPs used as a screening tool only.	Rotating basin monitoring done for assessment purposes, but also used for TMDL development. Current (second) cycle of monitoring is focused more on TMDL development monitoring.	No	Has used a probability-based program since the inception of the rotating basin monitoring strategy. This program provides statistically-based results for the target stream population of 1st - 5th order wadeable streams, and data are extrapolated per basin management unit, providing the commonwealth data to identify areas of concern, and eventually trends based on biological data. Not clear if probability-based data used for listing purposes.	Draft 303(d) list is published for public comment. Data from public monitroing groups will be utilized beginning with the 2008 305(b) cycle provided they meet QA/QC requirements. Data are obtained through a cooperative effort with the USACE, TVA and USFS when aquatic resources under said management exist in a basin management unit. The KY Division of Water also shares data with those interested agencies.	
Louisiana	Assessment depends on use and water quality parameter. Contact Rec use impairment based on 25% exceedance of fecal data, temp > 30% exceedance. Fish and Wildlife use: DO >10% exceedance, conventionals >30% exceedance, metals and toxics use 2 or more exceedances in most recent 3 years. Do not assess nutrients because no state standards for nutrients. Drinking water use: Fecals, Color >30% exceedance, metals and toxics use 2 or more exceedances in recent 3 years. All assessments use raw score.	Uses data from state's rotating basin monitoring program, as well as third-party data. Third-party data is requested and considered as part of 305(b)/303(d) process. Third-party data must meet QA/QC requirements established by LDEQ and be applicable to the water bodies being assessed during a given assessment cycle. Assessments based on data for DO, temp, pH, tubidity, fecals, chloride, sulfate, TDS and metals. Only data since 1997 considered. Generally minimum sample size of 5 for raw score assessments.	Sample collection, handling and analysis must be in accordance with state's EPA approved QAPP. No credible data law in effect.	If data are bad/rejected, they are not used at all (not for assessments, TMDLs, etc). However, data not meeting all QC requirements may still be usable, but will be flagged as estimates and usability determined.	Louisiana's fixed station ambient monitoring network is designed specifically for 305(b) and 303(d) purposes, however, data is used for other purposes as well.	No	No data extrapolation	Method developed using EPA guidance. Data solicited from public.	Good monitoring coverage of all waterbodies of the state through fixed station and rotating basin monitoring program.
Maine	Assessment based on type of criteria and beneficial use. Specific decision-rules are not provided for most parameters. BPJ referenced for assessments. Noted that single excursions or excursions within sampling error are disregarded.	Chemical, physical and biological data, including data from state probabilistic sampling, "qualitative" data from professional sources, modeling data, census data and indigenous species data. No minimum sample size.	Data used for assessment must meet Department approved QA/QC requirements. No credible data law.	Data not used for assessment purposes.	State uses rotating basin design for targeted sampling, and probabilistic monitoring designs where applicable, but network not specifically designed for 305(b) or 303(d).	Yes, for certain methods. Biomonitoring assessments incorporate a probabiity statement in the criteria. Other monitoring (e.g.fish tissue, bacteria) incorporates population sampling that provides statistical distribution information.	No data extrapolation	Third-party data solicited from public where there is approved QA/QC.	
Maryland	Assessment method differs per criteria type. i.e., Bioassessments use estimates of mean IBI and IFI compared to threshold, 10% raw score method for chemical data, "weight-of-evidence" approach using the Sediment Quality Triad for sediment data.	Uses all readily available. Generally uses MBSS data for biological assessments. Use quantified chemical data in water column, sediment and fish tissue, as well as sanitary surveys for bacteria and toxicity data. Minimum sample size of 10 for chemical data, if less available then case-by-case determination. Minimum of 10 samples for 8-digit watershed bioassessments, if less available then moves to 12-digit watershed bioassessment.	Non-MBSS bioassessment data divided into Tier 1 (used with MBSS) and Tier 2 (may augment assessment data) categories based on QA/QC and other factors. Will review QA/QC of third-party data and use BPJ to determine if data will be used in assessment. No credible data law.	BPJ will be used to determine if data will be used in assessment. May be used as a screening value to target further monitoring.	MBSS is probabilistic random sampling program focusing on 8-digit watersheds for bioassessment purposes for 305(b)/303(d). Other monitoring is tied to location of resource, level of recreational use, and/or to provide adequate state coverage.	Uses 90% confidence intervals on IBI and IFI scores when comparing to threshold. Bacteria monitoring uses a geometric mean, 90th percentile and/or single sample maximum. No uncertainty quantification with chemical data.	No data extrapolation	303(d) list submitted for public comment prior to finalization. Third-party data is solicited and used in assessment if determined adequate using BPJ.	MBSS monitoring network and bioassessment methodology provides statewide coverage and determination of watershed health. Well-documented bioassessment methodology supported by references.
Massachusetts	Assessment depends on use and parameter, using weight of evidence approach. Reference values in literature sometimes used for assessment rather than numerical criteria. Decision rules and BPJ provide the basis for assessment. Conventionals: 10% exceedance rule, toxicity tests: <75% survival. Biomonitoring used as final arbiter of borderline cases for aquatic life use. Fish consumption advisory = impaired. Drinking water: exceedance of MCL or more than conventional treatment required. Restricted shellfish harvesting = impaired. Recreation: exceedances of bacteria criteria, or swimming closures > 10% of the swimming season.	State will accept all readily available data, but to be used in assessment, must meet QA/QC requirements. Chemical, physical and biological data used.	Yes, data must have a QAPP, including lab QA/QC plan, use a state certified lab, have a data management QA/QC plan, and must provide documentation of QA/QC including analyses. Third-party data is used. No credible data law.	Not used. Evaluated data can be used to determine "support".	Rotating basin monitoring used for several different purposes, not specifically designed for 305(b) and 303(d).	No	No data extrapolation.	Public review of 303(d) list.	

State	How does method specifically determine Water Quality Standard attainment?	What specific type of data are required to conduct waterbody assessment method? Is there a minimum sample size?	Are there specific QA/QC requirements for data used in the assessment? Is there a credible data law in effect?	If data do not meet specific QA/QC requirements, how is the data used in the assessment?	How is the assessment method tied to monitoring network design (i.e. is there a specific monitoring program for 305(b) or 303(d))?	Does assessment method quantify uncertainty?	How does method extrapolate water quality over space/time where data are not available?	Was public participation/peer review sought when developing method? Is third-party data solicited from public?	Notable strengths of method
Michigan	For chemical constituents, nonattainment based on assessing the magnitude and frequency of exceedances of the WQS, based on a case-by-case basis. Also a "poor" rating of biological community (based on comparison of bio index to pre-defined threshold values) = impaired. CSO's = nonattainment. Bacteria exceedances of average criteria, or >10% of max criteria = impairment. Fish consumption advisories = impaired. Mercury and PCBs compared to criteria. Sediments use reference screening levels. Declining trend showing non-attainment by next cycle = impaired.	Uses state data and data from "credible" sources. Includes chemical, fish tissue and biological data, fish kills, modeled data. Only uses data from previous 5 years. Data must be representative of current conditions. No minimum sample size, except 5 samples required for fish tissue for mercury.	Yes, data are used in assessments that meet MDEQ QA/QC requirements, though the requirements are not specifically enumerated. No credible data law.	Data not meeting QA/QC requirements are used to justify further monitoring/assessment where the data indicate a problem.	Rotating basin monitoring program for surface water assessments and NPDES, not specific for 305(b) or 303(d).	No	Geographic extent of impaired waterbodies based on federal Reach File 3 indexing. Lakes considered impaired in entirety.	Draft methodology and 303(d) list submitted to public for comment.	
Minnesota	Assessment per pollutant category and use. Different assessment results generated depending on whether for 305(b) or 303(d). Impairment Examples: Toxics = 2 or more exceedances in 3 years. Conventionals = >10% exceedance. Fecals = >10% for > 2 months. Nutrients = exceedance of thresholds by ecoregion. Biological community =exceedance of basin-specific thresholds. Fish tissue = > 2ppb Hg or PCB. Final listing decisions based on "weight of evidence" and BPJ from a "professional judgement team" formed for each basin. Both quantitative and qualitative criteria are used for assessment of attainment of use.	Very specific data type and minimum sample size for each pollutant category assessment, includes chemical, biological and fish tissue data. Third-party data is solicited and used if approved by MPCA (meets data quality criteria). Different data requirements for 305(b) use vs. 303(d) use.	Specific criteria must be satisfied for use of third-party data, including QA/QC and appropriate monitoring objective of data. MPCA Quality Management Plan has been approved by EPA. No credible data law cited.	Not used in assessment, or if not enough data, waterbody not assessed and waterbody placed in Cat 3. If exceedance of criteria is indicated, then data is flagged for futured monitoring/investigation.	Only data from "condition" monitoring program is used in 305(b) and 303(d). This includes fixed-station (Milestone) stations for chemistry and rotating basin statistically-based (EMAP) stations for biomonitoring. Data from other monitoring programs (other information objectives) not used.	Not formally, but is recognized. MPCA's method of minimum sample sizes, screening data for exceedances, then applying BPJ results in a balance of the error of determining impairment when it isn't true, vs. not determining impairment when it is true.	Use EMAP to select bio-monitoring stations to represent watershed.	Public review of 303(d) list.	Very thorough methodology, though calcs are simple (not statistically rigorous). Formalized BPJ process. Recognition of relationship between 305(b) and 303(d) and clear explanation of how assessments are done for each. Don't use "targeted" sampling data in assessments.
Mississippi	Asessment differs per designated use. Biological community surveys preferred (i.e. given greater weight) for assessment of aquatic life use. Detailed biological assessment methodology given in Mississippi Benthic Index of Stream Quality (M-BISQ) program documentation. For chemical constituents, data are compared to state water quality criteria, only assess those for which numeric criteria have been developed. Single exceedance not used to list as impaired. BPJ used to assess impairment when only "slight" variation from criterion occurs. Can not be fully supporting of ALUS if no DO data. Use 10-percent rule for determining attainment for conventionals and toxics. Toxics only assessed against acute criteria when sampling does not match chronic criteria frequency (4-day avg). For rec use, impaired if >25% of samples violate geo mean std, or if >10% exceed fecal instantaneous criteria. Fish and shellfish consumption use based on advisories.	All readily available for most recent five years. Data must meet specific quality, quantity and sampling frequency requirements. Utilize chemical, biological, physical and toxicological data. For 2006 assessment of aquatic life use, used benthic macroinvertebrate data (required one survey in last 5 years). DO and other daily monitored conventional assessments require minimum of 3, 24-hr sampling events over 2-year period (during critical period) for daily average assessments, require minimum of 20 data pts over 5-yr period (no more than 10 pts in one year) during critical period for instantaneous minimum assessments. Toxics assessments require 10 data pts over 3-yr period. Bacteria (rec use) assessments require 4 sampling events over 2-yr period, for enterococci must have 20 samples (over 6 months) per event, for fecals must have 5 samples (over 30 days) per event.	Data need to be of highest quality and developed using sampling and analytical protocols and SOPs recognized by state and EPA QAPPs. No data will be assessed without supporting QA documentation. Benthic macro data collection must be compatible with MDEQ SOPs developed for M-BISQ program.	Data not meeting requirements can still be used in 305(b) assessments, or can be used for listing if data demonstrate compelling evidence of degradation and data is supported by data quality documentation. Waterbodies with less compelling evidence of impairment or lack of data quality documentation are placed on state's future monitoring list.	Fixed-station and rotating basin monitoring done for multiple purposes. M-BISQ monitoring done for 305(b) and 303(d) assessments.	No	No data extrapolation	Data solicited from public. Assessment methodology (CALM) published as part of 305(b) report.	
Missouri	Assessment based on use and type of data. Decision rules are established for attaining/non-attaining of use. Aquatic life: biological data shows "statistically significant" difference from reference streams = impaired, toxics acute >1 exceedance in 3 years, toxics chronic >10% exceedance, conventionals >10% exceedance. Other uses compare average of data to standard to determine attainment. Drinking water use: MCL exceedance = impaired. Also uses "observations" for compliance with narrative standards.	Uses "scientifically defensible" data for 303(d) listing purposes. State ranks data that is received for each waterbody as Level 1 - 4, depending on amount, age and type of data available. All data are considered, but only data considered "scientifically defensible" used to list on 303(d). No minimum sample size, though amount of data affects level designation.	State data is collected under an approved QAPP. Third-party data is reviewed by state to determine if satisfies QA/QC requirements. No credible data law.	Level 1 data is placed in Category 3 and used to prioritize waterbody for additional monitoring.	Several state monitoring networks developed for numerous purposes, not specific to 305(b) or 303(d).	Certain methods may quantify uncertainty, when statistics are used.	No data extrapolation	Draft methodology and 303(d) list submitted to public for comment. Data solicited from public.	Establishing levels for data quality for each waterbody (amount, age, type).
Montana	Assessment based on use. State has Record Sheet for each assessment documenting the data and analysis, and performs QA review on assessments as well. State also reviews/revises beneficial use support determinations at intervals not to exceed 5 years. For aquatic life: uses "weight-of-evidence" assessment when high level of data are available (physical, chemical & biological data available and 2 or more biological assemblages); uses "independent-evidence" assessment when less data are available. Other uses compare to numeric standards. For chemical constituents, uses raw score method (10% moderate, 25% severe), or compares data to surface water quality criteria. Also compares data to reference conditions (nutrients, sediment, habitat).	Gathers all readily available data, including chemical, physical and biological data, but only uses "sufficient credible" data for assessments and listing. At a minimum, must have at least 2 data types to be sufficient credible, unless "overwhelming evidence" of impairment (i.e. fish kills). These data may be either primary data (collected by DEQ for the purpose of WQ standards attainment assessment) or secondary data (collected by other agencies or organizations for a similar purpose).	Montana has law requiring that "sufficient credible data" be used in use-support determinations. For aquatic life use: SCD determination "scores" dataset by looking at technical soundness of analysis/methodology, spatial/temporal coverage, data quality and data currency.	Not used.	Montana has statewide fixed station monitoring network for multiple purposes, including 305(b). Also does targeted sampling for 303(d) in order to achieve sufficient credible data in a 1-3 day event for each waterbody.	No.	No data extrapolation.	Both assessment methodology and Integrated Report submitted to public for review. Assessment method originally submitted stand alone to public in 2000 for comment, as well as with each 303(d) list. Data is solicited from public via website, newspapers and flyers.	QA/QC of assessments, Assessment record sheets. Description of how reference conditions are established. Review/Revision of assessments at least every 5 years.
Nebraska	Assessment applies binomial method (90% conficence level) to determine "true" 10% exceedance for impairment listings. Exception: E. coli uses seasonal geometric mean compared to threshold; for DW, running annual average can not exceed MCL. Exceedances during low-flow (i.e. 7Q10, 1Q10 or 30Q5) do not count.	Uses all readily available data, quantitative, qualitative, and predictive. Data must be collected within last 5 years to be used in assessment. Water chemistry minimum sample size of 10. Assessment methodology geared towards chemical data.	QA/QC requirements must be met for data to be used. QA/QC and metadata should be submitted with third-party data. Data which exceeds holding times or is estimated is not used in assessments.	Data not verifiable or representative won't be used for 303(d), but may be used in planning future monitoring.	State has both fixed station and rotating-basin monitoring programs, though not specifically dedicated to 305(b) and 303(d) purposes.	Yes, 90% confidence level chosen for binomial procedure and allowable number of exceedances.	Data representative of waterbody unless major hydrologic change occurs. Stream stations can not exceed 25 miles, and lake stations can not exceed 25% of volume.	Data was publically solicited, and draft 303(d) list was submitted to public for comment.	Low flow considerations of criteria exceedances.

Appendix C. State Evaluation Matrix

State	How does method specifically determine Water Quality Standard attainment?	What specific type of data are required to conduct waterbody assessment method? Is there a minimum sample size?	Are there specific QA/QC requirements for data used in the assessment? Is there a credible data law in effect?	If data do not meet specific QA/QC requirements, how is the data used in the assessment?	How is the assessment method tied to monitoring network design (i.e. is there a specific monitoring program for 305(b) or 303(d))?	Does assessment method quantify uncertainty?	How does method extrapolate water quality over space/time where data are not available?	Was public participation/peer review sought when developing method? Is third-party data solicited from public?	Notable strengths of method
Nevada	Generally use 10% exceedance for previous 5 years to determine as impaired for 303(d). For aquatic life (acute and chronic constituents), waterbody listed if 2 or more exceedances of 1-hr criterion in 3 year "block" (if at least 10 samples). Single exceedances of Median/Average standards = impairment. Frequency, duration and magnitude of independent samples may warrant listing based on BPJ. Uses different assessment methodologies for 303(d) and 305(b), but moving toward integration. Assessments which show exceedance of "requirements to maintain existing higher quality" (i.e. antidegradation) not used to put on 303(d) list.	Uses all readily available data, including chemical, physical, sediment, fish tissue, biological, toxicity tests, narrative and qualitative info. Majority of assessments based on numeric data. All sample data from previous 5 years used for current assessment, including third-party data. Minimum sample size of 10, however, if fewer than 10 samples, more than 2 exceedances can warrant listing. Spatial and temporal representativeness based on BPJ. Data from extreme high/low events may be excluded as allowed in WQS. Biological data not used to list because bioassessment protocols not developed.	No specific QA/QC requirements, though data must have adequate documentation on data development and sample location. No credible data law.	Waterbodies not meeting minimum requirements placed on "List of Waterbodies Warranting Further Investigation".	State monitoring not specific for 305(b)/303(c) purposes.	10% raw score method accounts for some tolerance in local natural system function, sample collection and lab analysis. This recognizes that local natural variability functions within the remaining 90% compliance range.	If data sampled at a watershed's "control point" indicates impairment, then this is extended to entire reach. Tributaries can be included in listing if hydrologic connection indicates effects to beneficial uses.	Draft 303(d) list submitted to public for comment. Public notice to solicit third-party data.	
New Hampshire	Assessment depends on designated use. Uses weight-of-evidence approach for aquatic life and sediment. For waterbody to be fully supporting, set of "core parameters" must indicate full support. Exceedances of standards due to naturally-occuring conditions are not considered impairment. 10% exceedance rule used for most numeric criteria assessments with a 2 exceedance minimum. Also uses a "magnitude of exceedance criteria", if >2 samples exceed then impaired, regardless of 10% rule or sample size. For bacteria, use both a geo mean comparison, as well as single sample max comparison. Biological/Habitat assessment compared to threshold based on "bioregion". Toxics uses 10% rule for >10 samples, or >2 exceedances for <10 samples, or >2 exceedances for acute criteria. Restricted consumptions for fish and shellfish = not supporting.	Uses all readily available data, including third-party. Maximum data age of 5 years for streams, impoundments, estuaries and oceans, 10 years for lakes. Data includes chemical, physical, biological, habitat and toxicity. Biological data given more weight for aquatic life use. Minimum sample size of 10 samples to determine full support attainment for most numeric criteria. 10 sample size minimum does not need to be met for impairment determinations. Data below 7Q10 low flow in mixing zones not used for listing. Modeled data only used to determine as threatened, not impaired (still can be listed).	State documents data quality as low, fair, good and excellent, depending on QA/QC procedures and documentation.	"low" quality data not used in assessments, but can be used to screen for future monitoring. Any size datasets accepted and used in assessments.	Have a probabilistic sampling program for 305(b) purposes.	No	Homogenous "Assessment Units" defined by HUC-12 boundaries and water body types, water quality standards, known pollution sources, size restrictions, major land use changes, designated cold water fish spawning areas.	Requested comments on CALM for 2006 assessment. Continuous request for data from the public, plus a formal request for 2006 assessment.	Magnitude of exceedance criteria. In 2006, use of internal database that allows state to track the assessment status of every parameter/DU/AU, as well as the the degree of support, not-support, and insufficient information (which is broken into potentially attaining or potentially not attaining).
New Jersey	In 2006, assessment will focus on attainment of uses. A minimum suite of parameters have been identified for each use assessment. If sufficient data is not available, the use will not be assessed. Potential impairments based on 2 exceedences of the WQ criteria will be further evaluated in conjunction with data from other stations in the subwatershed, other related water quality data and previous status, to determine whether the waterbody should be listed as impaired based on BPJ. Recreation use is based on an exceedence of the geomean enterococcus in coastal recreational waters and E.coli in freshwaters. Uses NJIS score (EPA RBP method) for bioassessments: moderate or severe impairment = Cat 5. Bioassessment outweighs DO, pH or Temp assessments. Exceedances of chronic ALU criteria during high flows will not be used to list unless persists for 4 days. 2 or more exceedances of human health criteria = impaired for DW use.	Reviews all readily available data, but considers QA/QC, monitoring design, data age, sampling location info, data documentation and electronic data management in determining if data will be used in assessments. Data must be spatially referenced. For ALU, must have RBP data, plus temp and DO. If water chemistry is available, require 8 samples, data should be sampled at least quarterly, and data record must be at least 2 years in last 5 years. Datatsets less than 8, but greater than 4 samples use a modified assessment. For rec use, bacteria data need 5 samples over 30 days. Data taken below 7Q10 flow not used.	Yes, QAPPs or adherence to state's quality assurance guidelines, which includes use of a state-certified lab, are required. State also requests "citable" reports with data. Data must be electronically available.	Data that does not meet QA/QC requirements is not used for assessment purposes but may be used to identify areas with potential problems for additional monitoring.	Monitoring network is not specific to 305(b) and 303(d) purposes, however, focused monitoring may be needed to confirm impairment or delisting decisions.	No	Extrapolation of monitored waters based on stream order, size of watershed, land use, impoundments and sampling station type. Extrapolation also used to "estimate" use support in waters adjacent to monitored waters, based on land use, pollution sources and BPJ. These waters are not placed in Category 5. 2006 watershed delineation based on HUC 14.	Data is solicited from public, local groups, dischargers and other agencies. The Integrated list as an amendment to the Statewide Water Quality Management plan must be noticed for public comment. An opportunity for public comment on the Assessment Methods is also provided.	Detailed extrapolation methodology. Assessment based on designated use attainment, not just WQ criteria attainment.
New Mexico	Assessment differs per use and parameter. Use bioassessments/habitat assessments for aquatic life use (state working on biocriteria, in meantime are using reference conditions from Plafkin, 1989). Conventional chem and bacteria data uses 15% exceedance for >7 samples, or one exceedance if <7 samples. Toxics uses 1 exceedance in 3 years. WET test failure more than once in 3 years = impairment. Fish advisories indicate impairment due to mercury. Evaluated assessment (i.e. "observed affects") are noted in ADB but do not determine impairment. Use weight-of-evidence with multiple data types, considering data quality, defined by state in tables. Also rank bioassessments and habitat assessments from 1 (poor) to 4 (excellent) based on supportnig data, protocols, reference conditions, etc...to quantify confidence associated with assessment.	Use all readily available data, including chemical, physical, biological, habitat and toxicological data, provided data sampling and analysis prcedures meet QA/QC requirements. Put data in "levels" based on quality, temporal and spatial coverage. Make a point of assessing impairment and using data from low-flow (drought) periods, which is different from other states. Only use data from last five years, though also look at data up to ten years old for large rivers. Qualified (i.e. censored) data is used in assessments, as well as continuous data. No mimimum sample size.	QA/QC requirements are set forth in state's EPA approved QAPP. QA/QC documentation must be submitted along with data to ensure it complies with state QAPP. QA/QC considered in defining data quality level (1-4). No credible data law.	Data that does not meet QA/QC requirements is not used to determine impairments.	Monitoring network design is specific to 303(d) and 305(b) purposes.	No	Uses assessment units, of which sampling stations are assumed to be representative.	Integrated list published for public review and comment, as well as the Record of Decision (ROD) document which provides basis for listing and delisting decisions. Third-party data solicited from public every listing cycle. Assessment protocols reviewed by EPA Region 6, and are periodicaly submitted to public for comment when substantial changes have occurred (last public review for 2004 IR). Future parameter specific assessment protocols (i.e. temperature) will be developed by a workgroupd which includes non-agency participants (i.e. stakeholders).	Workgroup participation in assessment methodology development.
New York	Various assessment criteria per use. Generally uses 10% exceedance (raw score) method to designate as "suspected" and 25% as "known" impairment. 305(b) assessments feed directly into 303(d) list (not two separate processes).	Data includes use restriction orders, water chemistry, surrogate indicators and qualitative information. Use monitored data where available but supplemented with evaluated data. Data used generally collected within last 5 years.	Not mentioned	Not mentioned	Uses a 5-yr cycle rotating basin approach that focuses on previously unassessed, non-supporting waters and/or waters with data more than 5 years old, not specific for 305(b)/303(d) purposes.	Uncertainty is not quantified, but assessments showing impacts are characterized as "known", "suspected" or "possible" based on specific criteria.	No data extrapolation	Draft 303(d) list is published for public comment, in addition, public workshops and solicitation are part of 305(b) basin waterbody assessment updates.	305(b) assessments feed into 303(d) list, with documentation to link the two.

State	How does method specifically determine Water Quality Standard attainment?	What specific type of data are required to conduct waterbody assessment method? Is there a minimum sample size?	Are there specific QA/QC requirements for data used in the assessment? Is there a credible data law in effect?	If data do not meet specific QA/QC requirements, how is the data used in the assessment?	How is the assessment method tied to monitoring network design (i.e. is there a specific monitoring program for 305(b) or 303(d))?	Does assessment method quantify uncertainty?	How does method extrapolate water quality over space/time where data are not available?	Was public participation/peer review sought when developing method? Is third-party data solicited from public?	Notable strengths of method
North Carolina	Assessment method depends on use and water quality parameter. For aquatic life use: macroinvertebrate and fish biological data assessments use previously established criteria to assign a bioclassification (fair or poor = impaired, good-fair,good or excellent = supporting). Monthly ambient parameters with criteria are assessed using 10% exceedance rule.	State and approved third-party data used. Chemical, physical, biological data, as well as fish consumption, swimming advisories and shellfish harvesting growing area classifications, land use data and regional drinking water reports. A minimum of 10 samples over more than 1 year are needed for assessing chemical parameters. Only data within the 5-year data window are used for assessments. Monitored data assessments have higher degree of confidence than evaluated data.	State data collected under established and approved QA/QC procedures. Discharger coalition data under a state MOA meet same criteria as state data. All third-party data are screened for data quality and quantity and assigned Level 1, 2 or 3 per EPA guidance for 303(d).	Level 1 data used along with state data for assessments. Level 2 or 3 data used to ID pollution sources, limit spatial extrapolation, or point to need for additional sampling locations as state resources allow.	Monitoring not specific for 305(b)/303(d). Monitoring program uses a basinwide approach to assess water quality by watershed as resources allow. Specific monitoring is conducted to further assess 303(d) listed waters to develop TMDLs.	No (built into use of 10% exceedance rule).	If no data is available, the water is noted as "No Data". If data is considered to be inconclusive or insufficient, the water is noted as "Not Rated" on an evaluated basis until more complete information can be gathered. Supporting ratings may be extrapolated up tributaries where no problematic discharges or changes in land use occur, and may be on an evaluated or monitored basis depending on data available. Impaired ratings are not extrapolated and are based on monitoring data only.	Integrated report and 303(d) list are submitted to public for comment. Third-party data also solicited.	
North Dakota	Assessment depends on use and parameter. For aquatic life, uses weight-of-evidence approach looking at both biological and chemical data. For chemical data, generally uses 10-25% exceedance rule (raw score method). For biological data uses multi-metric index method and either compared to thresholds per ecoregion (one ecoregion uses reference values, other ecoregion data are pooled and 10% and 75% are used as thresholds between supporting, insufficient data, and not supporting). Bacteria data uses both geo mean standard 10% exceedance rule for individual samples. Lake assessments use both trophic status and qualitative information to determine if uses are not met. Fish tissue weighted-average mercury < 3 ppb.	Only uses "sufficient and credible data" defined as having documented QA/QC, data from recent 10 years (streams) or 12 years (lakes), minimum sample size of 10. Waterbody can be listed based on "overwhelming evidence" if minimum sample size not met.	Yes, documented QA/QC required for data to be used.	Not used	Not specific to 305(b)/303(d) purposes.	No	Assessment units defined by 8-unit HUC, same water quality standards classification, contiguous Level IV ecoregion, and stream order.	Yes, third-party data is solicited from public. Draft 303(d) list published for public comment.	
Ohio	Assessment depends on use and parameter - have a tiered designated use approach to water quality standards setting. For human health, use fish tissue concentrations compared to criteria. For recreation, pooled bacteria data and compared average and max criteria to 75th% and 90th% respectively. For aquatic life, compare fish and macroinvertebrate index scores to biocriteria in state WQ standards. Still drafting methodology for drinking water. Do not assess trends for "threatened" category.	Uses data from specific sources that state is confident of data quality, including fish consumption advisories, bacteria, and biological. Minimum data requirements specific to each use/parameter.	Ohio recently passed a credible data law requiring data to be "level 3" or higher. Because of this, Ohio did not solicit third-party data for 2004 IR. In 2006, only solicited bacteria data.	Not used.	Not specific to 305(b)/303(d) purposes.	No.	Uses watershed assessment units to apply results from biosampling to waterbody/stream segment.	State has a peer "advisory group" to develop recommendations. Also, Draft IR submitted to public for comment, and bacteria data solicited from public.	Tiered deisgnated uses and establsihment of biocriteria
Oklahoma	Assessment method differs per use, defined in states Continuous Planning Process (CPP). Uses the 10% rule as a default analysis protocol for short-term average parameters. Compares the 2-year rolling average to criteria for long-term average parameters. For attainment of some uses, must meet criteria for multiple chemical parameters or biological data, whereas violation of just one parameter results in non-attainment. Biocriteria have been set by the state for certain ecoregions, assessment based on using IBI scores compared to thresholds. If biocriteria not set, IBI must be at least 40% of regional reference condition. For fecals, compare geomean to criteria, or 25% of individual samples must comply with criteria. For E coli and Enterococci, one exceedance = non-attainment. Causes and sources are recorded in state's assessment database.	Use chemical, biological and physical data that meet data requirements. Stream data older than 5 years and lake data older than 10 years will not be used for attainment determinations unless no other data available. For streams, minimum of 10 samples for conventional and coliform. For large lakes, minimum of 20 samples. For toxics, a minimum of 5 samples. If existing data indicate impairment, minimum samples do not apply (i.e. for toxicant uses 2 order of magnitude above criteria, for DO uses 2 mg/L). CPP outlines specific non-detect handling procedures.	Data must meet minimum quantity, quality, temporal and spatial requirements. Data must be collected in accordance with EPA QAPP requirements, though documentation does not appear to be required.	Not used to determine use attainment.	Not specific to 305(b)/303(d) purposes.	No	Non-wadable stream monitoring sites may represent up to 25 miles, wadable sites up to 10 miles. Lake samples may represent up to 250 acres.	Listing methodology sent to interagency, EPA Region 6 and tribes for comment. Listing methodology is part of Continuing Planning Process, which has an extensive public participation. Public solicitation of water quality data for the Integrated report, including "nominations" for the 303(d) list, as well as public review and comment of draft Integrated report, including draft 303(d) list.	
Oregon	For 303(d) - specific assessment methodology per water quality parameter (chemical and physical) with numeric criteria. Some standards specify a mean or average over a specific time period. Other pollutants are determined to be "attaining" the criterion if 90% of 10 or more samples meet the criterion, or all samples in a small set of 5 to 9 samples meet the criterion. 305(b) summarizes available information from other surveys and projects conducted throughout the state which may be done using a variety of design and evaluation methodologies.	The assessment to determine attainment or non-attainment of standards is done on chemical and physical data over 10-yr period if available. Data requirements for minimum sample sizes are specified for each water quality parameter. Generally a minimum sample size of 5 samples is required to assess standard attainment unless the standard specifies other data requirements. Data is solicited from third parties for these assessments. The 305(b) report summarizes other available information from ambient monitoring sites, including results of an indexed score evaluation, and probabilistic sampling projects in high priority areas in the state.	Yes, only data meeting specific QA/QC requirements (Level A and B specified in Oregon's Assessment Methodology) are used for assessment of standard attainment. There is no specific "credible data law" in Oregon.	Data level C and E are not used in the assessment to determine standard attainment.	303(d) - uses data from 144 site large stream monitoring network, as well as other targeted monitoring programs (i.e. TMDL). No monitoring specific to 305(b).	No, not for standard attainment (303(d)). Some probabilistic survey data and analysis used in 305(b) provide confidence levels for results.	305(b) - probability-based survey designed so that stations represent larger watershed. 303(d) - use all available data to characterize waterbody. If only one site available then that site represents entire waterbody.	Draft Integrated Report and 303(d) list are submitted to public for review. Public notice for third-party data solicitation. Draft methodology is publicly reviewed during the "call for data".	Summary assessment sheets per parameter in 303(d) CALM.

Appendix C. State Evaluation Matrix

State	How does method specifically determine Water Quality Standard attainment?	What specific type of data are required to conduct waterbody assessment method? Is there a minimum sample size?	Are there specific QA/QC requirements for data used in the assessment? Is there a credible data law in effect?	If data do not meet specific QA/QC requirements, how is the data used in the assessment?	How is the assessment method tied to monitoring network design (i.e. is there a specific monitoring program for 305(b) or 303(d))?	Does assessment method quantify uncertainty?	How does method extrapolate water quality over space/time where data are not available?	Was public participation/peer review sought when developing method? Is third-party data solicited from public?	Notable strengths of method
Pennsylvania	Assessment depends on use and parameter. For bacteria data (rec use), two violation of geo mean standard = impairment. For aquatic life use assessments, primarily uses biological community assessments, impairment based on BPJ. For chemical data, state "compares" data to water quality standard.	Considers all readily available data, but must meet minimum data requirements. Biological, chemical, physical, habitat, fish tissue and bacteriological. Require minimum of 24 samples collected over 12-24 months in last five years for chemical/physical data. Stream segment definition required for all data types. For bacteria data, must have 5 samples over 30 days for at least two months.	Yes, third-party data must be accompanied by either a written study design, a written QAPP, or standard QA/QC protocol.	Data not used in assessment, but useful for public education or targeting waters for further study.	Specific monitoring network for habitat and benthic macroinverterbrate surveys exists and is described in the state's Comprehensive Monitoring Strategy but it serves more purposes than just 305(b)/303(d).	The chenical water quality assessment method and all biological assessment metrics employ benchmarks that consider the precision of the dataset used in their development, though 305(b) and 303(d) reports do not list quantified uncertainty.	Fixed stream segments used to characterize the spatial extent of assessment efforts are tied to the NHD. Trend analyses of water cemistry at fixed monitoring stations considers changes over time.	Method published for public comment with draft list. Third-party data solicited from public.	Rigorous minimum data requirements.
Rhode Island	Assessment method differs per use. Aquatic life: partially supporting if macroinvertebrate data show less than full support through modification of community and/or more than 1 exceedance in 3 years (but <10% of samples). Shellfishing: partial if an occasional or seasonal closure, not supporting if closed. Drinking Water Supply: fully supporting but threatened if supply appears to be threatened by non-point sources, or occasional detections of pollutants. Partial if one or more exceedances of MCLs, treatment beyond conventional. Not supporting if frequent exceedances and/or source closure due to contamination. Swimming: partial if geo mean standards of fecal coli met but >10% samples > 500MPN/100ml. Not supporting if geo mean not met. Fish consumption: Not supporting if "no consumption" order. Draft CALM guidance (2006/2007) may differ from above.	Chemical (water column grab samples) and biological. Use third-party data, often funds third-party monitoring efforts. Require minimum of 4 samples.	No specific QA/QC requirements for 2004 IR, though third-party QA/QC is reviewed. QA/QC requirements will be included in CALM guidance for 2008 IR. No credible data law.	Third-party data used because of limited state monitoring.	No specific monitoring network. Data is solicited. However, state to begin rotating basin approach and geometric sampling design in future.	No.	Single monitoring station is representative of waterbody both upstream and downstream where no significant influences exist. One lake sample at deepest location is representative of whole lake.	No public participation in past, but Draft CALM guidance will be issued for public comment in 2006-2007 for 2008 assessment cycle.	
South Carolina	Assessment based on use and parameter. For conventionals and bacteria, uses 10% exceedance rule (raw score method). For toxics, 2 or more exceedance in 5 years = impaired. For turbidity and nutrients, uses 25% exceedance rule (raw score). However, for all aquatic life and rec use parameters, will not list as impaired based on only 1 exceedance. Biological data (EPT Index and BI compared to thresholds) trump chemical data for aquatic life use. Fish consumption or shellfish harvesting advisories indicate non-support.	Solicits and considers all readily available data.	State has EPA approved Quality Assurance Management Plan - data are reviewed for adherence to State's QA/QC criteria.	Not used.	State has monitoring network, but not specifically for 305(b)/303(d).	No	No data extrapolation	Data solicited from public for 303(d) assessments.	
South Dakota	For monitored streams: >10% exceedance = not supporting, >25% if less than 20 samples. For lakes: Carlson's Trophic Status Indices developed for each lake and compared to ecoregion threshold (45-65). Also impaired if beach closures due to pollutant.	Chemical (water column) from state's Surface Water Quality Monitoring Program and others. Data obtained from STORET for assessment of each water segment.	State protocols have QA/QC requirements for state monitored data. Requires consensus that third-party data meets these QA/QC requirements as well. QA/QC info encouraged to be submitted with third-party data. No credible data law.	Not used in assessment.	State-wide fixed station ambient monitoring program for streams & Statewide lake assessment program from 1989-2003. However, this state-wide data is not the only data used in assessments.	No, though minimum sample size chosen so that >10% exceedance would be statistically significant.	No data extrapolation	Public input sought through third-party data solicition. Draft Integrated Report was also published for public comment.	Public comment solicited on Integrated Report.
Tennessee	Assessment appears to be both quantitative and qualitative, generally uses 10% exceedance rule to determine impairment if minimum data requirements are met. Also uses BPJ to consider season, history, rainfall, field observations, etc. to determine impairment. Single excursion can generate additional sampling for verification. Appears to favor biological assessments for aquatic life use assessments. Uses ecoregion reference approach to establish guidelines and compare data to local background levels.	Chemical (water column), bioassessment, habitat assessment data. Quasi-minimum sample size for some data types (4 samples), but with qualifiers can determine impairment with less. Third-party data solicited and used.	State has draft QAPP for their own monitoring program. State takes into consideration if QA/QC has been performed, but no specific requirements. No credible data law.	Data with questionable sampling techniques or analysis methods were not used in assessment, but are used to screen for future monitoring/studies.	State-wide intensive rotating basin monitoring program - used not only for Intergated Report, but other regulatory purposes.	No	No data extrapolation, but did recently break waters into smaller segments for assessment purposes.	Data was publically solicited, and draft 303(d) list was submitted to public for comment.	Rotating basin-monitoring approach to intensively monitor state-wide.
Texas	For numeric criteria, uses binomial method with 10% exceedance rate to determine impairment. Primary concerns are for water quality "indicators" (i.e. constituents) for which numeric criteria have been developed - these are assessed for 303(d). Secondary concerns are for indicators that do not have numeric criteria - these assessments used in 305(b), but not 303(d). Bacteria data compared to geo mean criteria, or 25% exceedance of single samples. For drinking water, long-term running average must be less than MCL. Fish consumption advisories = impairment. Biological data use Index of Biotic Integrity Scoring and compare to statewide or regional thresholds to determine use support. For aquatic life use, state developed decision matrix to integrate biological, chemical and habitat data assessments into final assessment.	State considers all readily available data, however data must meet quality requirements, no older than 5 years, and in a form not requiring extensive format manipulation. Appears that data in state database used for assessments, and third-party data used to verify analysis, or target future monitoring. Data older than 5 years can be used at state's discretion (i.e. trends). Data must be distributed over 2 seasons and 2 years (except for sediment and fish tissue). Generally a minimum of 10 samples to be listed, but not if data are over two year period and meet threshold of exceedances. Drinking water data requires 4 samples. Biological and habitat assessments require 2 datasets.	Yes, minimum QA/QC requirements include documentation of QA/QC procedures.	If data doesn't meet QA/QC not used in statistical analysis, but can be used for planning and identifying water quality concerns. Data not meeting minimum sample sizes not used for assessments.	Statewide monitoring not specific for 305(b)/303(d), though targeted monitoring conducted to strengthen assessments.	Through use of binomial method. Number of exceedances chosen to keep Type I error <= 20%, except for acute criteria and small sample size (<20), where Type I error chosen to be <= 50%.	Spatial extent of data points based on BPJ, though cannot represent more than 25 miles. Data from stream segments less than 25 miles, or lake less than 8 square miles,may be aggregated to meet 10 sample minimum.	Data solicited from public and draft 303(d) list submitted to public for comment.	Decision matrix for aquatic life use integrating biological, chemical and habitat assessments. Recognition of difference in assessments for data with numeric criteria vs. not.

State	How does method specifically determine Water Quality Standard attainment?	What specific type of data are required to conduct waterbody assessment method? Is there a minimum sample size?	Are there specific QA/QC requirements for data used in the assessment? Is there a credible data law in effect?	If data do not meet specific QA/QC requirements, how is the data used in the assessment?	How is the assessment method tied to monitoring network design (i.e. is there a specific monitoring program for 305(b) or 303(d))?	Does assessment method quantify uncertainty?	How does method extrapolate water quality over space/time where data are not available?	Was public participation/peer review sought when developing method? Is third-party data solicited from public?	Notable strengths of method
Utah	Assessment depends on use. For drinking water, 2 or more violations or >10% exceedance (raw score method) = impaired. 303(d) list includes any stream segment with NPDES discharge that does not have technology-based or end-of-pipe effluent limits. For bacteria (primary and seconday contact), uses geo mean standard or 25% exceedance of single value standard to determine impairment. Conventionals use 10% exceedance rule. Toxics use 2 or more exceedances in 3 years or 10% exceedance rule for acute, 3 or more exceedances in 3 years for chronic (also looks at "overwhelming evidence" by multiplying standard by 1.75 for less than 10 samples. Additional BPJ used for some aquatic life support assessments (trophic state index, fish kills, algae, biological/habitat data).	Considers readily available data, including third-party. Generally use chemical data for assessments, though biological and habitat data were used on a limited basis to supplement chemical data for aquatic life use assessments. Minimum sample size of 4 for acute toxics, 10 for chronic toxics.	Not mentioned	NA	State has rotaing basin monitoring network for 305(b) and other purposes	No	Used 10-digit and 12-digit HUCs to delineate stream AUs	Data solicited from public. 303(d) list reviewed by technical advisory committee and public.	
Vermont	Relies mainly on bioassessments (comparing biological integrity to state biocriteria) for aquatic life use. Uses stream geomorphic assessment for physical characteristics/habitat (compares to reference condition). Toxics: acute >1 exceedance in 3 years = impairment, chronic >4 days in 3 years = impairment. Bacteria: exceedance of geo mean standard = impairment. State also relies on statistical hypothesis tests for assessments, using different tests for different types of data. Frequency of exceedances to determine use support done on case-by-case basis (consistent failure or exceedances). Tests and hypothesis are chosen to maximize power and minimize Type I error (typically chosen to be 5 or 10%).	Considers all readily available data, including biological, physical, habitat, geomorphic, chemical, algal blooms. Minimum data requirements for bacteria data (5 samples over swimming season, at least 2 seasons). Bioassessment require a minimum of 2 assessments with findings of "fair" or lower quality to list a waterbody as impaired for ALU, unless finding of poor or very poor (only one assessment needed).	Data generated in conjunction with QAPP are used in assessment. Third-party data requires assurance that QAPP has been followed before used in assessment.	Data of unknown quality given less weight of other, higher quality data are available, or used to target waterbodies for more monitoring.	State monitoring networks are specifically designed around 305(b)/303(d) data needs.	Yes, through selection of Type I error rate in hypothesis tests.	No data extrapolation beyond that of sampling design.	Listing methodology published for public comment. Data solicited from public.	Use of statistics and bioassessment methodologies
Virginia	Method differs per criteria and use - very detailed list of type of data and criteria for placing waters into various categories and subcategories. For example, conventional parameters: at least 2 exceedances and >10.5% exceedance = impaired. Aquatic life - 30-day SPMD parameter exceeds criterion 2 or more times in 3 years. Trend analysis indicating exceedance of criteria during next listing cycle = impaired.	Accepts all readily available data, though classify information as "QA/QC approved monitored" data or "evaluated" data. Third-party data solicited and used, depending on data quality and subsequent classification. No minimum sample size, though sample size does impact which category water is assigned (no listing on one sample).	Very specific QA/QC and metadata requirements for data to be considered "QA/QC approved monitored data". No credible data law.	All data used to classify waters, but only approved data used for 303(d) listing category.	Large Water Quality Monitoring Program that does support Integrated Report, but has other data objectives as well. State recognizes the diverse functions of their monitoring program and categorizes sampling stations with a designation of the specific objective of that monitoring site.	No	No data extrapolation.	Yes, water quality assessment guidance is public-noticed prior to publication of Integrated Report. Additionally, state requires University scientists to be involved with development of assessment guidance.	Very detailed assessment methodology and use of subcategories so that all data can be considered, though may not be used for 303(d) listing. WQ data liason to work with public, etc to share data.
Washington	305(b): Do specific assessments per use using the 10% and 25% criterion exceedance thresholds. Also do a statistical inference assessment for overall use support to apply to whole state. 303(d) Assessment differs per criteria. For conventional pollutants: 10% exceedance per binomial method at 90% confidence. Sediments: site-specific (no numeric stds). Toxics: 2 or more exceedances in 3 years, or average of 3 highest single fish (or one composite fish) tissue toxics exceeds EPA threshold. Bacteria: Fecals geo mean exceed geo mean std, or >10% of samples used to calculate geo mean. Temp & DO: at least one 7-day avg exceeds std. Narrative: site-specific documentation of alteration, compared to reference, including how this impairs the use, and if it is human-caused. Also site-specific info on impairment due to flow (Category 4, not 5).	305(b): Use state data exclusively. 303(d): Appear to use all readily available data, as data is solicited from all interested parties. However, assessment methods focus on quantified data. Data must meet QA/QC requirements to be used in assessment. Data must have minimum metadata to be used in assessment (date, location, pollutant, value and units). Minimum samples: toxics (2 in 3 years), fecals (5), others (3).	303(d) - Yes, a QA/QC plan is required for data submitted, and in 2006, QAPP will have to be provided along with data submittal. Certain minimum metadata are required for use. A credible data law was recently passed and credible data policy is in effect for 2006 listing cycle.	Data won't be used to place in Category 5, but could be placed in Category 3.	Have a routine ambient network, and an EMAP stream network. Data is used from both of these for 305(b), though they are not designed exclusively for 305(b) or 303(d).	Yes, 90% confidence level for binomial method for conventional pollutant assessments.	For 305(b), do a statistical proportion estimate to determine status of use support over entire state. No extrapolation for 303(d), as this is considered inappropriate for 303(d) scope and purpose. Data extrapolation left to TMDL studies as needed.	State presents an Integrated Report format list (i.e. "Water Quality Assessment") to the public. The WQA includes Category 5 listings (aka the 303(d) list) as well as all other categories (which represent the 305(b) report). After public comment the WQA is sent to EPA for approval of the Category 5 portion. EPA had access to to the entire WQA, but approves only the "303(d)" portion.	Flow impairment is recognized, though not used for 303(d) listings. The presentation of the WQA to the public on the Internet using tools (Simple Query Tool and Interactive Mapping Tool) specifically designed to select and portray waters and the locations of impairments makes the information much more accessible and understandable to the general public.
West Virginia	Assessment depends on use and criteria parameter. For aquatic life, 2 or more exceedances of acute numeric standard in any 3 yr period = impairment. For chronic aquatic and human health criteria, >10% exceedance rule (raw score method) for 20 or more samples or >3 exceedances for <20 samples = impairment. Exceptions (not listed) if most recent data show no exceedances. Bacteria based on comparison of monthly geo mean to geo mean criterion (minimum 5 samples per month) or 10% exceedance rule for comparing single samples to daily criterion. Benthic macroinvertebrate metrics compare community to reference and "score" waterbody (score of 60.6 or less are impaired). 2 meals/month or more restrictive fish consumption advisory = impaired. Mandatory parameters must show attainment to be placed in Category 1 (full support), bacteria for contact rec use and benthic macroinvertebrate for ALU.	Uses all readily available chemical and biological data, generally from previous 5 years. Do not use data from low flows (<7Q10) or mixing zones. Minimum sample size of 20 generally needed to apply 10% exceedance rule, with alternative procedures for limited datasets. Minimum of 5 samples over 30 days for bacteria geo mean assessment.	State reviews data to make sure collection, analysis and QA/QC consistent with approved procedures.	Not used.	Not specific to 305(b)/303(d) purposes.	No	If data indicate impairment, waterbody is considered impaired upstream and downstream until another sampling location indicates nonimpaired condition. Historical impairment determinations retained until contradicted by new information of similar strength.	Solicited data from public. Draft 303(d) list submitted to public for comment.	Mandatory parameters to designate as "fully supporting"

State	How does method specifically determine Water Quality Standard attainment?	What specific type of data are required to conduct waterbody assessment method? Is there a minimum sample size?	Are there specific QA/QC requirements for data used in the assessment? Is there a credible data law in effect?	If data do not meet specific QA/QC requirements, how is the data used in the assessment?	How is the assessment method tied to monitoring network design (i.e. is there a specific monitoring program for 305(b) or 303(d))?	Does assessment method quantify uncertainty?	How does method extrapolate water quality over space/time where data are not available?	Was public participation/peer review sought when developing method? Is third-party data solicited from public?	Notable strengths of method
Wisconsin	Appears that any excedance of criteria = impairment. Also unique aquatic life procedure determines impairment if the "existing" use is less than the "codified" use (i.e. aquatic life species present). Must have two separate declining trends to determine as "threatened".	Generally uses state data, but also solicits third-party data, as long as it meets the criteria for "monitored" data. Uses Waterbody Assessment and Reporting System (WADRS) as tabular and spatial database to organize data for 305(b) and 303(d) assessments.	To be classified as "monitored" data, must be site-specific, representative of current conditions, and have "adequate" QA/QC. No credible data law.	Considered "evaluated" and not used for 303(d).	Comprehensive Baseline monitoring program is set-up to accomplish information needs of 305(b). Focus on aquatic life & habitat. Designed to answer if designated uses are being met, status and trends. Includes wadeable (random stratified), non-wadable, lakes, fixed station (for trends). Also special studies monitoring.	No	No data extrapolation	Through solicitation of data, and public review of 303(d) list.	WADRS geo-referenced database of data used for assessment and reporting.
Wyoming	Assessment based on weight-of-evidence approach. For numeric/chemical criteria, uses 10% exceedance rule, but also considers magnitude of exceedance (if less than 10 samples). For bacteria, compare to geo mean std, but also list if one sample is >double the standard. For aquatic life use, acute or chronic criteria exceeded more than once in 3 years = impaired, or declining trend in water quality suggesting future exceedances. Also considers "modification of aquatic community beyond natural range of reference conditions".	For numeric/chemical criteria, only requires that type of data. For narrative criteria, uses "credible" data as defined by state law, i.e. scientifically valid chemical, physical and biological monitoring data collected under an accepted SAP including QA/QC procedures. State also considers ancillary data (soil, landuse, climate, etc). Minimum of 5 samples over 30 days for bacteria.	Yes, QA/QC procedures required under credible data law.	Not used.	Not specific to 305(b)/303(d) purposes.	No	Some data extrapolated over space, based on watershed characteristics and land uses.	Draft 303(d) list and methodology submitted to public for comment.	

Glossary of Terms

(Adapted from U.S. EPA, 2005a)

Assessment Unit (AU) The waterbody segment whose attainment status is reported in the Integrated Report. An AU may be designated based on homogenous hydrologic characteristics (i.e., stream order), designated uses, land cover, land use, and locations of point and non-point pollution sources.

Attainment The condition of meeting applicable water quality standards.

Assessment Making a decision about the condition of a segment with regard to water quality standards.

Cause(s) of Impairment The stressor(s) (i.e., pollutant), whose presence in a waterbody is causing or contributing to failure to meet any applicable water quality standards.

Data Quality Objectives A specification of the quality of the data needed in order to meet the monitoring project's goals.

Designated Uses Those uses specified in state water quality standards regulations for a particular segment, whether or not they are being attained. Uses so designated in water quality standards are not meant to specify those activities or processes that the waterbody is currently able to fully support. Rather, they are the uses/processes that the state wishes the waterbody to be clean enough to support, whether or not the waterbody can, in its current condition, fully support them.

Exceedance A situation in which ambient conditions are inconsistent with those desired conditions described by the water quality criteria.

Excursion Having a concentration that is inconsistent with the concentration specified by the water quality criteria.

Impairment Failure to support a water quality standard.

Indicator A specified pollutant, or other chemical/physical/biological condition, such as phosphorus, copper, temperature, pH, *E. coli*, etc.

Quality Assurance Project Plan (QAPP) A written document that outlines the procedures a monitoring project will use to ensure that the samples participants collect and analyze, the data they store and manage, and the reports they write are of high enough quality to meet project needs.

Quality Assurance (QA) The overall management system which includes organization, planning, data collection, quality control, documentation, evaluation, and reporting activities. QA provides the information needed to ascertain the quality of data and whether or not they meet the requirements of a project. QA ensures data will meet defined standards of quality with a stated level of confidence.

Quality Control (QC) Routine technical activities whose purposes are, essentially, error control. Since errors can occur in the field, laboratory or office, QC must be part of each of these functions.

Population The set of individuals being studied.

Sample A single measurement or aliquot.

Segment A waterbody (e.g., river, lake, bay, estuary, wetland) or portion thereof.

Waterbody Assessment (see Assessment)

Water Quality Criteria (Criterion) Elements of state water quality standards, expressed as indicator levels representing a quality of water that supports one or more designated uses. Numeric water quality criteria addressing chemical, physical, or biological conditions should contain three attributes: 1) magnitude, 2) frequency, and 3) duration. *Magnitude* is that element of a water quality criterion specifying acceptable levels of a pollutant or other indicator. Most criterion magnitudes are expressed as concentrations. *Frequency* describes how often waterbody conditions can exceed the magnitude, and *duration* describes the averaging period over which ambient data is averaged for comparison to the criterion magnitude.

Water Quality Standard Provisions of a state law which define the water quality goals for a waterbody/segment. Water quality standards consist of designated uses, water quality criteria, antidegradation policies and implementation procedures.

References

Arizona Department of Environmental Quality (ADEQ). 2004. *The status of water quality in Arizona – 2004: Arizona's 2004 Integrated 305(b) Assessment and 303(d) Listing Report.* Water Quality Division.

Carlson, R.E. 1977. A Trophic State Index for Lakes. *Limnol. Oceanogr*. 22(2): 361-369.

Federal Water Quality Coalition (FWQC). 2002. *Preparation of Integrated Water Quality Monitoring and Assessment Reports: Recommendations for Clean Water Act 303(d) and 305(b) Methodologies and Reporting*. Prepared by Barnes and Thornburg LLP, Parsons Corporation. Published by the Federal Water Quality Coalition, American Farm Bureau Federation and the Association of Metropolitan Sewerage Agencies.

General Accounting Office (GAO). 2000. *Water Quality – Key U.S. EPA and State Decisions are Limited by Inconsistent and Incomplete Data*. GAO/RCED-00-54. U.S. General Accounting Office, Washington D.C.

Gibbons R. 2000. *Alternative Statistical Approaches for Performing Water Quality Impairment Assessments Under the TMDL Program.*

Griffith, L.M., R.C. Ward, G.B. McBride, and J.C. Loftis. 2001. *Data Analysis Considerations in Producing 'Comparable' Information for Water Quality Management Purposes*. National Water Quality Monitoring Council Technical Report 01-01. http://water.usgs.gov/wicp/acwi/monitoring/pubs/tr/

Helsel, D.R. 2005a. More than Obvious: Better Methods for Interpreting Nondetect Data. *Environmental Science and Technology*, October 2005: 419-423.

Helsel, D.R., 2005b. *Nondetects And Data Analysis: Statistics for censored environmental data.* Wiley, Hoboken, New Jersey. 250 p.

Helsel, D.R. and R.M. Hirsch. 2002. *Statistical Methods in Water Resources,* USGS Techniques of Water Resources Investigations, Book 4, Chapter A3, 510 p. Available online at: http://water.usgs.gov/pubs/twri/twri4a3/

Iowa Department of Natural Resources. 2005. *Methodology for Iowa's 2004 water quality assessment, listing and reporting pursuant to Sections 305(b) and 303(d) of the Federal Clean Water Act.* TMDL & Water Quality Assessment Section, Geological Survey & Land Quality Bureau, Environmental Services Division.

Karr, J.R. and C.O. Yoder. 2004. Biological Assessment and Criteria Improve Total Maximum Daily Load Decision Making. *Journal of Environmental Engineering*, June 2004: 594-604.

McBride, G.B. 1999. Equivalence tests can enhance environmental science and management. *Australia & New Zealand Journal of Statistics*, 41(1):19-29.

McBride, G.B. 1998. Statistical methods: when differences are equivalent. *Water & Atmosphere*, 6(4):21-23. National Institute of Water and Atmospheric Research, New Zealand.

McBride, G.B., J.C. Loftis, and N.C. Adkins. 1993. What do significance tests really tell us about the environment? *Environmental Management*, 17(4): 423-432. (Errata in 18(1):317).

Miller, D. 2006. U.S. U.S. EPA Region 7 Water, Wetlands and Pesticide Division. Personal Communication.

National Research Council (NRC). 2001. *Assessing the TMDL Approach to Water Quality Management*. National Academy Press. Washington, D.C. www.books.nap.edu/html/tmdl/

Public Employees for Environmental Responsibility (PEER). 1999. *Murky Waters: Official Water Quality Reports are All Wet, an Inside Look at U.S. EPA's Implementation of the Clean Water Act*. www.peer.org/execsum.html.

Smith, E.P., K. Ye, C. Hughes, and L. Shabman. 2001. Statistical Assessment of Violations of Water Quality Standards Under Section 303(d) of the Clean Water Act. *Environmental Science and Technology* 35: 606-612.

SWAQ. 2003. Subcommittee for Water Availability and Quality, President's Office of Science and Technology Policy. http://water.usgs.gov/owq/swaq.pdf

U.S. EPA. 1990. *Biological Criteria, National Program Guidance for Surface Waters*. U.S. EPA-440/5-90-004. Office of Water Regulations and Standards: United States Environmental Protection Agency, Washington, D.C.

U.S. EPA. 1991. *Policy on the Use of Biological Assessments and Criteria in the Water Quality Program.* Office of Science and Technology: United States Environmental Protection Agency, Washington, D.C.

U.S. EPA. 1997. *Guidelines for Preparation of the Comprehensive State Water Quality Assessments (305(b) Reports and Electronic Updates: U.S. EPA 841-B-97-002A and U.S. EPA 841-B-97-002B*. Office of Water Regulations and Standards: United States Environmental Protection Agency, Washington, D.C.

U.S. EPA. 2001. *U.S. EPA Requirements for Quality Assurance Project Plans.* U.S. EPA QA/R-5. United States Environmental Protection Agency, Washington, D.C.

U.S. EPA. 2002. *Consolidated Assessment and Listing Methodology: Towards a Compendium of Best Practices, First Edition*. U.S. EPA Office of Wetlands, Oceans and Watersheds: United States Environmental Protection Agency, Washington, D.C.

U.S. EPA. 2003a. *Guidance for 2004 Assessment, Listing and Reporting Requirements Pursuant to Sections 303(d) and 305(b) of the Clean Water Act*. Watershed Branch: Assessment and Watershed Protection Division: Office of Wetlands, Oceans and Watersheds: United States Environmental Protection Agency, Washington, D.C.

U.S. EPA. 2003b. *Elements of a State Water Monitoring and Assessment Program*. U.S. EPA 841-B-03-003. Office of Wetlands, Oceans, and Watershed: United States Environmental Protection Agency, Washington, D.C.

U.S. EPA. 2003c. *A guide to developing secondary information products: methods review and documentation*. U.S. 260-B-01-006. Office of Information Analysis and Access: United States Environmental Protection Agency, Washington, D.C.

U.S. EPA. 2005a. *Guidance for 2006 Assessment, Listing and Reporting Requirements Pursuant to Sections 303(d), 305(b) and 314 of the Clean Water Act*. Watershed Branch: Assessment and Watershed Protection Division: Office of Wetlands, Oceans and Watersheds: United States Environmental Protection Agency, Washington, D.C.

U.S. EPA. 2005b. *Guidelines for Carcinogen Risk Assessment*. Risk Assessment Forum: United States Environmental Protection Agency, Washington, D.C.

U.S. EPA. 2006a. *Guidance on Systematic Planning using the Data Quality Objectives Process* U.S. EPA QA/G-4. United States Environmental Protection Agency, Washington, D.C.

U.S. EPA. 2006b. *Data Quality Assessment: A Reviewer's Guide U.S. EPA QA/G-9R*. United States Environmental Protection Agency, Washington, D.C.

U.S. EPA. 2006c. *Data Quality Assessment: Statistical Methods for Practitioners*. U.S. EPA QA/G-9S. United States Environmental Protection Agency, Washington, D.C.

U.S. EPA. 2006d. Water Quality Standards: Basic Information. URL: http://www.epa.gov/OST/standards/about/. Last updated on May 11, 2006.

WERF. 2003. *Navigating the TMDL Process: Listing and Delisting*. Stock No. 00WSN2. www.werf.org

Yoder, C.O. 1997. *Important Concepts and Elements of an Adequate State Watershed Monitoring and Assessment Program*. ASIWPCA Standards and Monitoring, Ohio U.S. EPA Technical Bulletin. MAS/1997-7-1. Columbus, OH: Ohio U.S. EPA Division of Surface Water.

Yoder, C.O. 2004. *Region V State Bioassessment and Ambient Monitoring Programs: Initial Evaluation and Review*. Midwest Biodiversity Institute & Center for Applied Bioassessment and Biocriteria, Cleveland, OH.

IWA Publishing's authorised EU representative for General Product Safety Regulations is Diane D'Arras, 15 rue Duret, 75116 Paris, France, e-mail: safety@iwap.co.uk.

Printed and bound by CPI Group (UK) Ltd, Croydon, CR0 4YY
12/05/2026
02108865-0010